ATM Interworking in
Broadband Wireless Applications

For a complete listing of the *Artech House Telecommunications Library*, turn to the back of this book.

ATM Interworking in Broadband Wireless Applications

M. Sreetharan
S. Subramaniam

Artech House
Boston • London
www.artechhouse.com

Library of Congress Cataloging-in-Publication Data
Sreetharan, M.
    ATM interworking in broadband wireless applications / M. Sreetharan, S. Subramaniam.
      p.   cm. — (Artech House telecommunications library)
    ISBN 1-58053-285-3 (alk. paper)
    1. Asynchronous transfer mode.   2. Broadband communication systems.
    3. Wireless communication systems.   I. Subramaniam, S.   II. Title   III. Series.
TK5105.35 .S84   2002
621.382'16—dc21   2002027953

British Library Cataloguing in Publication Data
Sreetharan, Muthuthamby
ATM interworking in broadband wireless applications. — (Artech House
telecommunications library)
    1. Asynchronous transfer mode   2. Broadband communication systems
    3. Wireless communication systems I. Title   II. Subramaniam, Suresh, 1968–
621.3'8216

ISBN 1-58053-285-3

Cover design by Ykatarina Ratner

International Standard Book Number: 1-58053-285-3
Library of Congress Catalog Card Number: 2002027953

10 9 8 7 6 5 4 3 2 1

To my friend from school days, aeronautical engineer V. Sornalingam
(nom de guerre, Lt. Col. Shanker) who, with his two brothers, gave his life
defending the freedom and dignity of the people of Thamil Eelam.

—M. Sreetharan

To my sister Sivagini, who succumbed to breast cancer at the young age of 37.

—S. Subramaniam

Contents

Preface

The convergence of data and voice networks is accelerating the need to use ATM technology in transport networks. In addition, broadband wireless technologies, such as LMDS, are increasingly being deployed between customer-edge and provider-edge equipment. This book attempts to provide a comprehensive understanding of the systems engineering principles and details needed to implement wireless broadband applications using ATM interworking methods with emphasis on traffic management and quality-of-service (QoS) issues. The importance in developing interworking arrangements between other data networking protocols is stressed.

The book begins with an overview of protocols and network interworking, addressing each subject in depth, and concludes with detailed traffic-management analysis, emphasizing broadband wireless systems. ATM borrows many ideas from Frame Relay and other communication technologies. The book provides extensive background on these wide area protocols.

Chapter 1 introduces networking relating to packet- and circuit-switched networks, wireless broadband, and the need for ATM interworking.

Chapter 2 covers in detail the IP, Frame Relay, and ATM protocols, which form the basis of networks today. It provides an extensive background of these technologies and introduces the standards organizations.

Chapter 3 covers the wireless component that defines LMDS. It begins with an overview of LMDS, MMDS, Point-to-Point, and WLL systems and describes the basic components of a typical LMDS system, covering frequency bands and modulation techniques, along with the two wireless access methods known as frequency division duplex (FDD) and time division duplex (TDD). It concludes with backhaul connectivity using optical standards.

Chapter 4 describes various legacy WAN protocols, such as Frame Relay, that need ATM interworking support for the transport of application-specific data. The popular Internet Protocol (IP) and Data Link Switching as access protocols are also described. With the deployment of each new protocol, the need for interworking becomes important. The physical media available for interfacing with ATM networks also plays a major role when deploying ATM. Due to the wide deployment of DSL for broadband services, ATM over DSL is also presented in the final section.

Chapter 5 covers the IP and LAN interworking standards that have evolved to support interworking with ATM. The introduction of Multiprotocol over ATM (MPOA) to support the interworking of non-IP protocols is described. The chapter concludes with the interworking of Signaling System Number 7 (SS7), used widely in circuit switching.

Chapter 6 covers the interworking of two wide area networking (WAN) protocols: Frame Relay and ATM. Frame Relay is widely deployed and the success of ATM depends on the interworking of these two protocols. The Frame Relay and ATM Forums have worked hard to provide appropriate standards for the interworking of ATM and Frame Relay.

Chapter 7 describes the emulation of time-division multiplexed (TDM) circuits known as circuit emulation service (CES). CES is widely used for voice emulation using 56/64K PCM.

Chapter 8 covers the traffic-management aspects of ATM, ATM service categories, and QoS. It focuses on the service agreement that defines the QoS and various leaky bucket algorithms. This chapter also describes ATM policing and shaping with respect to user traffic.

Chapter 9 covers the practical aspects of implementing ATM in LMDS. Issues with ATM cell clumping in a typical TDMA-based LMDS are discussed. Dynamic bandwidth allocation (DBA), which allows the efficient use of LMDS bandwidth, is described in detail with an application example. In the LMDS example described in this chapter, the attachment of CES, Native ATM, Frame Relay, IP (LAN), and DSL interfaces to a point-to-multipoint configuration gives details of bandwidth and traffic requirements used by the various types of ATM service layer categories. The delay requirements for CES and interworking issues relating to wireless are also discussed.

The book concludes with Chapter 10, which assesses the impact of emerging technologies, such as 3G wireless, on ATM interworking.

Acknowledgments

We'd like to thank everyone involved in the Point-to-Multipoint (PMP) Program at Hughes Network Systems (HNS), Germantown facility, especially Dan Wendling and Bhanu Durvasula, who gave us the opportunity to participate in the PMP project. We learned a lot from our engineering colleagues at HNS who worked on this project. Our special thanks to them and to Mike Lohman, Bob Fischler, Laura Lafaver, and Tayyab Khan for architecting a successful PMP product.

We would like to thank Bill Highsmith, Systems Engineer, P-COM, Inc., who gave us permission to use material from his white paper, "Converging Complementary Technologies to Maximize Service Flexibility," which is used as the main background material for Chapter 3, and Dona Ruth Bjorson, who was responsible for creating all of the artwork. We are also grateful to Pratheev Sreetharan for the attention and care he took in generating Figure 1.1, and to Kuga Kugarajah for helping us with the editing.

We thank our reviewer for his detailed comments and criticism, which contributed to quality of the technical content and organization of the book. The authors accept full responsibility, however, for any remaining errors and omissions. We are grateful also to Barbara Lovenvirth, our Artech editorial assistant, for without her friendly reminders on slipping deadlines, the book would have remained in draft form for several more months.

To my talented friends at PCI—Ed Crane, Peter Springston, Amy Brongo Newman, and others—who have been a rich source of motivation and support, I am grateful. Finally, I am grateful to my sons, Rajeev, Pratheev, and Sanjeev, and to my wife, Mathini, for enduring long hours without protest while I struggled with the book after work.

Muthuthamby Sreetharan

I'd like to thank my wife, Sumathini for her encouragement during the long hours of working on the manuscript and my children Sasika and Ganesh for their humor and teasing about the manuscript and some of its unusual acronyms (e.g., ATM—automatic teller machine) that kept me going.

Sivanasnda Subramaniam

1

Introduction

Wireless networks for data and voice communications systems are establishing themselves as important components of traditionally fully hardwired networks. Initially popularized by the mobility they offered to data and voice users, wireless networks are increasingly being deployed as competition against wire-line and optical leased line provisions to businesses, infrastructure (cellular backhaul) applications, and as local loop for new entrants to the telecommunications market. These networks alleviate costs involved in hardwiring communication lines to user premises and allow fast deployment. Wireless systems have also facilitated provision of new data/voice services to multitenant units (MTUs) and to businesses in built-up city centers in shorter deployment times, thus increasing the competitiveness of businesses using wireless facilities. Figure 1.1 illustrates a point-to-multipoint (PMP) wireless network deployed in an urban environment.

While edges of voice and data network experience the fast deployment of the wireless component, asynchronous transfer mode (ATM) technology is increasingly becoming the central technology in the workgroup and enterprise network environments and as the transport technology of choice in nationwide wide-area networks (WANs). ATM provides scalable bandwidths and quality-of-service (QoS) guarantees at attractive price performance points, facilitating a wide class of applications that can be supported in a single network. The bandwidth-hungry applications spawned by the World Wide Web (WWW), as well as the general use of the Internet, telephony,

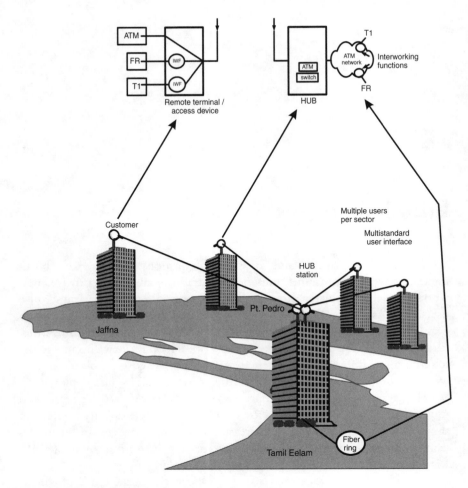

Figure 1.1 PMP wireless system in an urban setting.

video-on-demand (VOD), and videoconferencing have given rise to the need for simultaneously supporting the services on the same network. ATM is serving as the main catalytic technology in promoting the convergence of multiprotocol multiple networks into a single network providing multiple services.

ATM is designed to meet the requirements of both network service providers and end users. With ATM, service providers and end users can establish priorities based on the real-time nature of the traffic. Delay-sensitive voice and real-time video traffic are often given the highest priority, and nondelay sensitive traffic such as e-mail and local-area network (LAN) traffic are given a lower priority. These priority levels allow the service providers to charge according to the QoS.

Because of ATM's popularity and its potential for providing orders of magnitude of additional bandwidth for user traffic, traditionally popular networks are being forced to use ATM in internetworking arrangements. Such arrangements and standards that allow Internet Protocol (IP) networks and Frame Relay networks to coexist with ATM provide the current deployers of these networks the flexibility to evolve networks toward the most cost-effective solution. Many such interconnected network arrangements already exist, and product offerings facilitating such arrangements are available today. Systems Network Architecture (SNA) and X.25 networks have internetworking arrangements to expand their life span in the predominantly IP, Frame Relay, and ATM networking infrastructure.

Another important trend is the move toward voice systems to use packet networks. As the rate of growth of packet data networks began to exceed that of the circuit-switched telephony networks, the need to deploy packet networks to carry voice has overtaken the need to build circuit-switched networks to carry data. Voice-to-data network interworking [and the need to carry Voice over Frame Relay (VoFR), IP, and ATM networks] is increasingly important to the convergence of data and voice networks into an all packet network.

Convergence to an all ATM network requires that proper consideration be given to maintaining the same or better QoS that users expect from ATM networks while the data is transported via non-ATM networks. The QoS aspect is further complicated when a wireless segment is introduced at the edge of the ATM network. Air-protocols may often alter the traffic pattern of the ATM cells that are carried over it. Impact on QoS arising from such situations must be addressed in the design of wireless systems in broadband interworking environments.

This book focuses on the following technical aspects of the use of ATM technology in wireless broadband networks:

As current data networks evolve into ATM networks, and as voice and packet data networks converge toward a single multiservice network, the development of interworking solutions becomes critical. Different access methods, and the required interworking need for using IP, Frame Relay, and, ultimately, ATM as the transport network, are identified and detailed. Interworking functionality that is currently employed, and the necessary interworking arrangements between the access protocols and the transport protocols will be described with detailed reference to authoritative compliance standards developed by organizations that are key players in the networking field. This book will also deal with how the mapping of traffic parameters between the components in an interworking arrangement is implemented to maintain the QoS that the network user expects the service provider to deliver.

Transition of the wireless component from an expensive segment into an integral part of the voice/data network will also be explored, and the performance impact on end-to-end voice/data sessions will be assessed. The effect of traffic parameters on QoS aspects in different protocol segments of the network will be detailed, and practical solutions on setting traffic parameters to accommodate the wireless component will be discussed. ATM will be used as the main network protocol in these discussions.

1.1   Need for Broadband and ATM

The International Telecommunications Union (ITU) has defined broadband as any rate higher than primary rate (T1 or 1.544 Mbps). Bandwidth demands of advanced multimedia applications built on the WWW, as well as the Internet, telephony, videoconferencing, and similar networking applications are at last being matched by advances in transmission technology and their application to networks.

Different characteristics of applications drive the need for broadband network support. They include the following:

- High traffic volumes necessitated by increasing use of traditional applications. Increasing use of e-mails that contain documents, graphics, and even moving pictures. Client-server architecture-based applications and large file transfers result in large volumes of traffic, which benefit from broadband networks. Increasing capacity-access networks, especially different forms of digital subscriber line (DSL) circuits, are providing low-cost, high-bandwidth access to single users to their homes, vastly increasing the traffic volumes the supporting transport networks need to carry.

- Special client-server applications, such as those supported by the WWW, are interactive in nature. Increased bandwidth is needed to provide acceptable latency for increasingly complex graphic-intensive contents or to support the increasing use of Web-initiated downloads of large files (such as executables or large audio-video files).

- Applications that are inherently broadband, such as moving digital images, digital photography, music, and VOD, necessitate a broadband infrastructure.

- Convergence of telephony and data networks, at the core transport level transmitting transparently circuit-switched data, or in the use of voice-over-packet technologies in packet-switching networks, will result in the need to transfer large volumes of voice/data necessitating increasing deployment of broadband networks.

Audio bit-rates range from CD-quality sound, which yields a 1.411 Mbps at 44.1 KHz sampling with 16-bit quantization, to 64 Kbps for digital telephony. With compression algorithms, MPEG layer 3 produces CD-quality voice stream of 128 Kbps with 12:1 compression rates, and typical vocoders produce 8-Kbps compressed speech rates in telephone networks.

Video streams produced by MPEG-1 to support 720 pixels sample density per line, 576 lines per frame, and 30 frames per second TV sizes have variable rate bit streams ranging from 1.86 Mbps to a maximum of 15 Mbps.

The mix of different services, in addition to having different bit rates, have different traffic characteristics and real-time needs that require the network to guarantee a defined QoS generally defined in terms of sustained bit rate, excess bit rate, and burst size.

To handle such traffic volumes and QoS guarantees, emerging networks have used latest transmission, switching, and protocol technologies. While narrowband transmissions generally operate over copper wires or coaxial cables, broadband transmission technologies use the large bandwidth provided by optical fibers. Currently, the wireless transmission technologies have also been developed to carry DS3 rates (45 Mbps) and above to form important segments of access networks in a broadband architecture.

Broadband rates are described in optical carrier (OC) rates:

OC-1	51.84 Mbps
OC-3	155.52 Mbps
OC-12	622.08 Mbps
OC-24	1244.16 Mbps
OC-48	2488.32 Mbps
OC-192	9953.28 Mbps

Advances in optical transmission technologies have recently produced dense wavelength division multiplexing (DWDM) technology. DWDM allows splitting of light into multiple wavelengths (colors), each of which can carry data at 10 Gbps and enables a single fiber to carry up to 100 different wavelengths giving an aggregate bandwidth of more than 1 Tbps per fiber.

Similar to T1 and T3 framing in traditional data/voice networks, the Synchronous Optical Network (SONET) standard, as part of a larger telephony standard called Synchronous Digital Hierarchy (SDH) by the Comite Consulatif Internationale de Telegraphie et Telephonie (CCITT), defines the framing for transmission of digital data over fiber. The basic building block in SONET is a synchronous transport signal Level 1 (STS-1) frame, which is organized as a 9-row by 90-column byte array, transmitted row first. The basic frame timing used in T1 circuits (i.e., 8,000 frames per second) gives an STS-1 rate of 51.84 Mbps ($90 \times 9 \times 8 \times 8,000$). SONET framing allows multiplexing of low-speed digital signals such as DS1 or DS3 and is able to integrate services such as ATM, which is made up of fixed length cells of 53 bytes in length.

ATM is a technology linked to the development of broadband Integrated Services Digital Network (ISDN) in the 1980s. As a packet-switched high-performance technology, ATM can support multiservice applications, including multimedia. Established standards for network and user interface, signaling, and traffic management have facilitated ATM's rapid growth and adaptation to different networking needs. ATM architecture provides switched and permanent categories of virtual circuit connections [i.e., permanent virtual circuits (PVCs) and switched virtual circuits (SVCs)] between end systems and promotes optimized utilization of bandwidth by defining different classes of services.

The explosive growth in the broadband needs of emerging applications dictate the need for fiber-based infrastructure development. The ATM networking technology and fiber synergy has encouraged network service providers to increasingly deploy ATM-based transport networks.

The following sections provide a broad overview of the technologies and protocols that provide the framework for the detailed coverage on ATM interworking.

1.2   Current State of Data and Voice Networking

Voice and data networks can be broadly grouped into three categories: Carrier networks of interexchange carriers (IXCs), enterprise networks, and small-scale networks at small businesses and homes. In addition, the incumbent local exchange carriers (ILECs), competitive local exchange carriers (CLECs), and Internet service providers (ISPs) provide access to end users to data and voice networks, completing the infrastructure needs for a wide variety of data and voice networking.

From the early days of enterprise networking, where discrete devices were connected to mainframe systems, data networking methods at enterprises have used different high-bandwidth LANs to connect hundreds of powerful workstations and servers. Enterprise networks and small-scale users employ their own private networks or use services offered by public packet-switched data network services that employ protocols such as Frame Relay, IP, and ATM.

1.2.1 Circuit-Switched Networks

From the first 600 point-to-point private lines that provided telephone facilities in 1877, the telephony network of today has been based on circuit-switched technology. Frequency division multiplex (FDM) techniques used in early analog systems gave way to digital transmission with the advent of T1 carrier systems in the 1940s. Time division multiplexing (TDM), which uses the DS0 (64 Kbps) as the basic building block carrying a single voice channel, is the dominant technology in the common carriers' fully digital networks. Digital multiplexing allows transmission speeds ranging from DS1 (1.544 Mbps, 24 voice channels) to DS4 (274.176 Mbps, 4,032 voice channels).

Central office switches (Level 5), which provide line interfaces to analog plain old telephone service (POTS) lines, dominate the network edge interface access devices. Switches of Level 4 and lower, the tandem switches, mainly perform trunk switching from T1 levels to increasing trunk capacity levels extending to T4.

SONET ring networks at the core provide a fault-tolerant broadband infrastructure for multiplexing digital voice channels to higher levels defined in the OC standards. The SONET format provides direct multiplexing of synchronous and asynchronous services into its STS Level N payload. Subrate multiplexing is accomplished by defining virtual tributaries (VTs) in SONET payload format. This feature at a physical level facilitates both data and voice networks to share the same optical medium with the use of add/drop multiplexers, which can insert and remove payloads originating from ATM networks, as well as DS1 to DS3 clear channel payloads from voice networks.

1.2.2 Packet-Switched Networks

In circuit switching, dedicated bandwidth is allocated to each user session and the end-to-end connection, from the beginning of a session setup (data or voice) to the end of the session, remains static. In contrast, packet switching allows sharing of network resources by making use of logical sessions between end users on a mesh network where data packets from different sessions are multiplexed on physical links connecting different network nodes. The statistically bursty traffic characteristic from different logical sessions is used to obtain optimal use of network resources.

Packet-switching technology gained popularity with the X.25 networks of the 1980s. Figure 1.2 illustrates the overall architecture of a typical packet network with user nodes data terminating equipment (DTE) and network nodes data communicating equipment (DCE). The Open System Interconnection (OSI) reference model (see Figure 1.3) developed by the International Standards Organization (ISO) defines the layered architecture that is used to describe the different layers of the communication protocols that support packet-switching networks. The key point to note in the layered architecture is that the transport layer and above reside only in the end systems; the network nodes that function as packet switches only have a

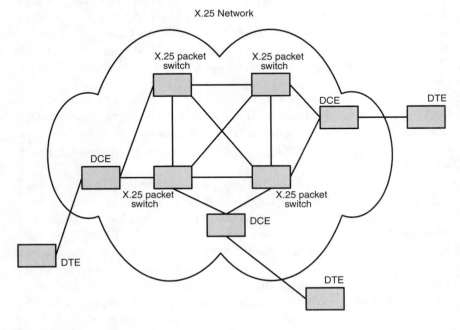

Figure 1.2 Architecture of a packet network.

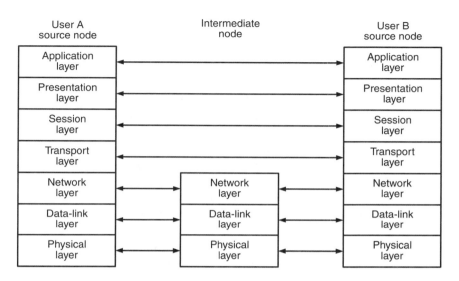

Figure 1.3   OSI reference model.

network layer and below, routing packets across the network, except for network management of the node, for which there are other application layers.

Packet-switching technology has undergone changes driven by the many-fold increase in speed of transmission media and the virtually error-free transmission of digital data over long distances. The protocols and network architecture has evolved to take advantage of the speed, reliability, and cost benefits afforded by the new technologies.

1.2.2.1   IP Networks

Currently popular IP networks originated in the U.S. Defense Agency's TCP/IP layer of ARPANET in the 1970s. The networks' development and widespread use are driven by the open nature of the Internet Engineering Task Force (IETF), a voluntary organization. Request for comments (RFC) documents generated by committees comprised of experts from industry and academia serve as the foundation for the continuing evolution of IP networks.

As opposed to the connection-oriented approach of the X.25 type networks, IP networks are based on the datagram concept where end-to-end reliable delivery of IP packets are not guaranteed by the IP layer. With flexible and sophisticated dynamic routing protocols such as the Routing Information Protocol (RIP) and the Open Shortest Path First (OSPF) in IP networks, packets can be routed around points of congestion and points of failure. The burden of sequencing, windowing, and reliable delivery is on the layer 4 protocol TCP that resides in the end systems.

The Internet represents the seamless interconnection of IP networks or IP servers owned by several users. The IP networks of different organizations are connected to the Internet by regional service providers, or ISPs. Multiple competing ISPs present in any region provide value-added services to their subscribers and manage the links to their customers. High-capacity links from the ISPs will connect to "transit" (backbone) providers to complete the global Internet infrastructure.

Interworking of IP-based networks with ATM will be one of the main areas of study in this book. The popularity of the Internet, giving rise to an unprecedented new generation of applications based on the WWW infra-structure, and, more recently, the integration of mobile browsers with the Wireless Application Protocol (WAP) ensure that IP will continue to be an important technology in new-generation networks. The IP version 6 (IPv6), where an enhanced IP packet format has been created to accommodate the dwindling IP address space, will introduce additional interoperability complexities that system developers and router vendors need to take into account.

1.2.2.2　Frame and Cell Relay Networks

Relaying variable-length frames (level 2 in the OSI model) and fixed-length cells are the basic concepts behind the frame and cell relay networks, respectively. The key to these new technologies has been reduced levels of processing at switching nodes and increasing levels of hardware participation in the switching algorithms, giving rise to increasing levels of session multiplexing and vastly minimized end-to-end propagation delays to match the many-fold increases that have been achieved by technology advances in the transmission technology.

Frame Relay, which represents an interface, a protocol, and a service, uses a connection-oriented service at the frame level with the use of PVCs and, less commonly, SVCs. Using multi-DS0 slots up to full T1-level speeds and allowing multiple connections to be carried over a physical link, Frame Relay networks evolved as the next-generation of networks to carry X.25 and IP traffic efficiently. Therefore, interworking agreements and standards of different protocols with Frame Relay are a precursor to the latest generation of networks employing ATM technology.

Figure 1.4 provides an overview of the architecture of current data net-work architecture, which uses the public FR/ATM/IP networks as transport

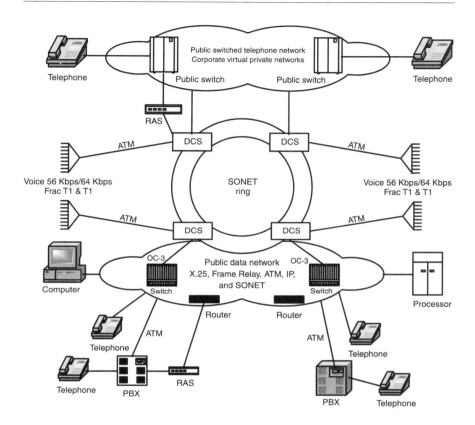

Figure 1.4 Architecture of a general data network.

networks. These networks have already started evolving toward accommo-
dating packet-switched voice.

Today's networks often consist of a large number of disparate architectures and protocols: LANs, both large and small with different configurations and different LAN protocols; ATM-to-desktop solutions represented
by campus networks that include ATM switching hubs and LAN emulation
(LANE) to facilitate interoperability among local ATM networks and LANs;
and large enterprise networks and hard-to-die legacy networks. It is not difficult to envision the multitudes of interworking functions intrinsic to the
proper operation of the different protocol segments of such networks. Further, the complexity of interconnections where a user's data traverses makes
it difficult to maintain the QoS promised to the end user. An example is
where data is first transported from the user's IP network via a Frame Relay

access device, and later by an ATM network to a destination IP server. In another example, adding a wireless segment when the Frame Relay data is piped into an ATM network in a local multipoint distribution system (LDMS) will burden the network designers in their quest to deliver guaranteed QoS to users.

1.2.3   Convergence of IP and PSTN Networks

Public switched telephone networks (PSTN) and IP networks are evolving toward a single packetized network that can handle voice and data. The convergence of data and voice networks will be a recurring theme in future carrier networks. In this scenario, ATM networks will be a major player in the core transport sections of the converged network. Although many of the technology issues described in this section do not directly relate to ATM interworking, they will, however, be components of future carrier networks where ATM will have a significant presence. Understanding the interworking aspects between voice and data will therefore be useful in developing future networks. QoS issues will present interesting and demanding challenges at interworking points where traffic parameters of one side of the network will need to be translated to an appropriate form relevant to the other side of the network.

Figures 1.5(a–c) provide a conceptual framework that illustrates the different possible high-level architectures that can play a role in the evolving convergence of networks and services.

In a functional framework, both the PSTN and the Internet can be viewed as consisting of a control part and a transport part. In a PSTN, these functions are often performed in different physical units. Signaling System 7 (SS7) subsystem components implement the control part, and the digital transmission systems are responsible for the transport part. In the Internet, however, the different layers of the protocol stack are responsible for these functions, the layer 3 and management applications are responsible for the control and lower layers, and the physical transmission systems perform the transport functions. The architecture in Figure 1.5(a) represents the most general and flexible form of convergence, where a gateway node in each path acts to translate the signaling and data segments across the two network types. Figure 1.5(b) refers to scenarios where routing and signaling for the PSTN is implemented via IP network facilities. The PSTN/Internet Interfaces (PINT) Working Group of the IETF has standardized the connection arrangements through which Internet applications can request PSTN services. The PSTN/IN Requesting Internet Service (SPIRITS) is another IETF

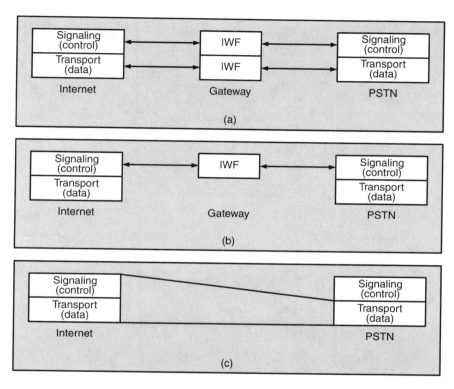

Figures 1.5 High-level functional layers in data/voice networks: (a) most generic form of convergence; (b) routing and signaling in PSTN done via IP; (c) portion on Internet overlayed in IP transport.

committee that is continuing to work on standardizing this type of service convergence. Figure 1.5(c) represents the common practical case where a portion of the Internet is overlaid on the transport segment of the PSTN. Dial-up access to the Internet and virtual private network (VPN) using tunneling represents some examples of this mode of convergence.

The different architectures and interworking issues that are employed in a flexible scheme depicted in Figure 1.5(a) is the concept behind Voice over IP (VoIP) telephony.

The architecture described in Figure 1.5(b) is the framework for the logical class of networks labeled intelligent networks (IN), where PSTN services are created, accessed, and managed by hosts in an IP network. Figure 1.6 illustrates the different components in such an arrangement. Internet call waiting (ICW) and different PINT services, such as click-to-dial-back

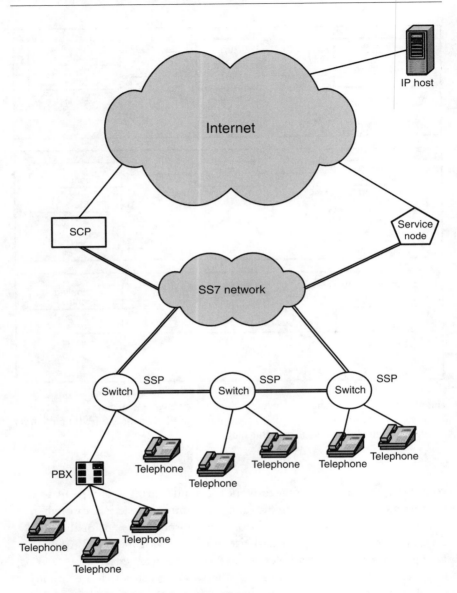

Figure 1.6 Components of an intelligent network.

described in RFC 2458, are examples where IP support is used to enhance PSTN services.

Physical access provisions to IP networks and tunneling schemes used in VPNs are examples of systems represented in Figure 1.5(c). The last mile plays a critical role for end users as new techniques in modulation and signal processing have evolved to extract increasing levels of performance from copper wires extending from the central office to the customer's premises. With dial-up modems reaching speeds of 56 Kbps and more, constrained by the 4-KHz bandwidth allocated to voice, DSL technology that uses the full bandwidth of the copper wire can reach transmission speeds of more than 10 Mbps. On these physical transmission layers, frame-based protocols such as Point-to-Point Protocol (PPP), Ethernet frames carrying IP packets, or cell-based protocols such as ATM will be carried to complete the logical connectivity with carrier networks.

Data carried over PSTN lines poses a challenge to PSTN Level 5 switches at central offices. Data connections are likely to stay alive 10 to 100 times or more than an average voice call, which lasts 3 to 5 minutes, and therefore would violate erlang loads that are used to scale the switch for local telephone traffic. Techniques to off-load Internet traffic by using proxy SS7 gateways or special modules in the central office switch to divert the data traffic to the ISP's access server are also becoming popular.

Traditionally dedicated lines or Frame Relay PVCs were used to connect islands of networks between enterprise branch offices. These were costly and difficult to manage when operational requirements dictated adding new lines or adding connectivity to new sites. Another class of applications where increasing the number of mobile users dial into remote access servers and modem banks to access corporate network to conduct their business were costly to set up and to manage. Replacing such systems are VPNs, where a shared infrastructure such as the Internet is used to transfer data between corporate sites and the ISP provides the virtual connections. Tunneling techniques, where packets with addresses have global significance and traverse two border points in the network in the encapsulated form, are key to implementing VPNs.

Figure 1.7(a) shows the difference between a direct dial and a VPN-based call, and Figure 1.7(b) illustrates the tunneling setup with a remote access server. It is worth noting that the Point-to-Point Tunneling Protocol (PPTP), the layer 2 Forwarding (L2F) protocol from Cisco (RFC 2341), and the Layer 2 Tunneling Protocol (L2TP) described in RFC 2661, were developed to support dial-up VPNs and operate at layer 2 of the OSI reference model, whereas IPSec, which functions as part of layer 3, is used mainly in LAN-to-LAN VPNs.

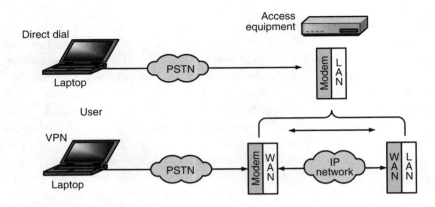

Figure 1.7(a)   VPN versus direct dialing.

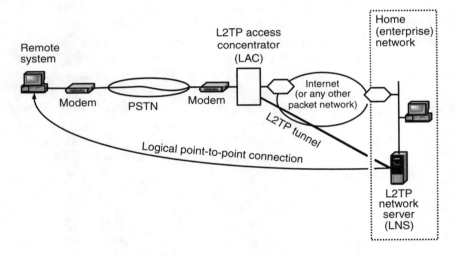

Figure 1.7(b)   Tunneling architecture.

1.3   Wireless Segment in Data and Voice Networks

Wireless networks have become important segments of data and voice networks. The aspects of wireless data transfer protocols that impact the QoS characteristics of data transfer sessions will be covered in this book. Broadband wireless systems, which can carry data at speeds ranging from fractional T1 to higher transfer rates, are more relevant to the current study than the wireless systems that support low data rates. The characteristics and use of the wireless component in a voice or data network differ widely and can be broadly categorized into four groups: wireless networks for mobile telephony, wireless cellular networks for data applications, fixed wireless networks, and broadband wireless access networks.

1.3.1 Wireless Networks for Mobile Telephony

These networks typically use 20-KHz-wide channels (for analog mobile phones or AMPs) and are used to connect mobile telephone users via base transeiver stations (BTSs) and base station controllers (BSCs) to a mobile telephone switch in the carrier's premises, which provides connection into the PSTN network. The general special mobile (GSM) is widely used in Europe and works with 20-KHz channels in the frequency range of 800 to 900 MHz. The General Packet Radio Services (GPRS) platform enables always-on wireless IP. The Universal Mobile Telecommunication Services (UMTS) platform is the latest platform attempting to provide cost-effective, high-capacity mobile communications with rates up to 2 Mbps under stationary conditions, with global roaming and other advanced capabilities. ATM is used to carry voice/data traffic in UMTS. It also plays a role as the integration platform for GSM, GPRS, and UMTS on a common, multiservice core network infrastructure.

1.3.2 Wireless Cellular Networks for Data Applications

Wireless telephony networks are also being used to send data. On completion of call setup, signaling data modems modulate the carrier with digital data and transmit in analog form similar to voice. The receiving modem demodulates the carrier to recover the transmitted data. Alternatively, data can be sent digitally using protocols the mobile telephony standard bodies have developed. GPRS is used as a data transfer protocol in a GSM-based telephony system. Cellular digital packet data (CDPD) technology is used to send data in carrier networks that support the AMPS system. The highest data rate CDPD supports is 19.2 Kbps. It has to be noted that CDPD is a digital technology that operates as an overlay network in the analog standards-based AMPS system.

1.3.3 Fixed Wireless Networks

The operational knowledge acquired in building mobile telephony infrastructure for countries with advanced economies enabled carriers to adapt the wireless segments for local loops. Fast installation, as opposed to the convenience of mobile access, was the driving factor for the emergence of fixed wireless networks. Similar to data services supported by the mobile cellular wireless networks, fixed wireless networks provide similar services. With the restricted bandwidth, these networks still provide users data capability such as dial-up wireless access to the Internet.

1.3.4   Broadband Wireless Access Networks

Figures 1.8(a–d) illustrate different configurations where users or enterprises access broadband networks with the aid of integrated access devices (IADs). The IADs serve as integrated bandwidth optimizing devices by combining information streams from different service interfaces ranging from Ethernet LANs and digitized voice from private branch exchanges (PBXs) on fractional T1s, to higher band campus ATM (25 Mbps) networks, aggregating the data streams to transmit at transport speeds from T1/E1 to optical carrier Level 3 or 12 (OC-3c or OC-12). Time division multiplexing (TDM) inputs from T1 can be a structured service where the IADs can perform add-drop multiplexing functions or the full T1 can be viewed as an unstructured service, which the IAD can pass transparently to the network via an ATM cell stream, to be reconstituted as a full-T1 at another exit point of the network.

Wireless access networks can replace the high-bandwidth consolidated access link of the IAD by a suitably configured high-capacity wireless link. Broadband wireless systems operating above 20 GHz (depends on the

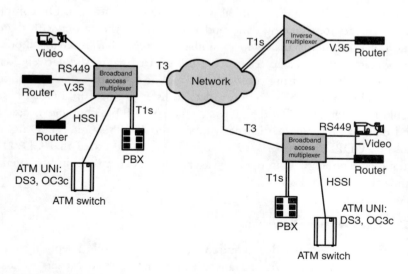

Figure 1.8(a) Broadband access multiplexer.

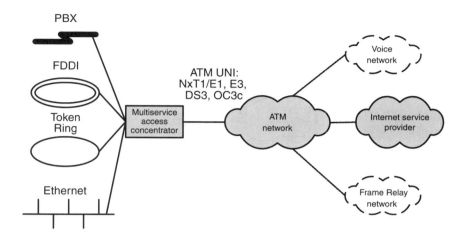

Figure 1.8(b) Edge concentrator.

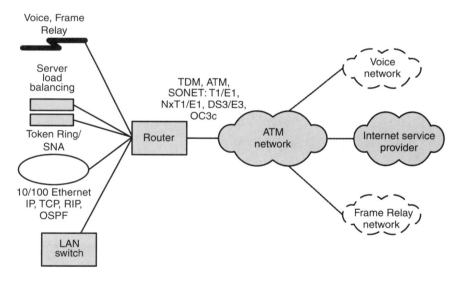

Figure 1.8(c) Broadband access router configuration.

country where licensing is issued) that use a PMP architecture are known as LMDS and have begun to penetrate the user access market at the periphery of high-bandwidth broadband networks that serve as transport nets. Spectrum ranges that are designated for broadband wireless services allowing two-way PMP communications services are

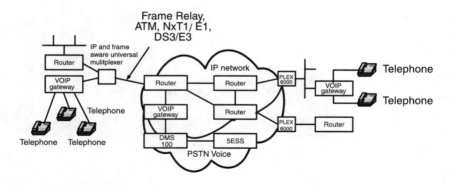

Figure 1.8(d) Intelligent or universal multiplexer.

DEMS: Digital Electronic Messaging Service

24.25–24.45 GHZ and 25.05–25.25 GHz

400 MHz in 5 of 40 MHz channels

LMDS: Local Multipoint Distribution System

28 and 31 GHz

A block: Total of 1,150 MHz

B block: 150 MHz in two 75 MHz blocks

38 GHz: 39.6–40.0 GHz

14,100 MHz blocks of 50 MHz each

In Chapters 3 and 9, we will concentrate on studying the characteristics of a typical LMDS system and how the wireless architecture introduces issues that affect QoS guarantees that have been given to a user of the service. Even in simple cases where native ATM cells are transported over the wireless medium, due to the characteristics of the air link protocol, the wireless system is unlikely to reproduce the traffic characteristics of the input ATM cell stream in its aggregated output stream. A typical air-protocol will be outlined to highlight these issues and design suggestions as to how QoS guarantees can be preserved across the wireless medium.

These issues are also present in low-bandwidth wireless applications, as interactive applications based on TCP/IP-based end systems are likely candidates for using such systems. However, since these users are generally used to the best-effort delivery characteristics of current IP nets, it is unlikely that they will expect guaranteed QoS in such situations.

Figures 1.9 and 1.10 provide a high-level view of the wireless component in a broadband network supporting voice and data. The wireless segment can be viewed as representing a practical LMDS service. The access node in the diagrams represents the remote terminal (RT) where multiprotocol ports are provided for user access and where interworking functions convert different incoming protocols to ATM. The network hub is the LMDS providers' hub terminal (HT), which connects to a broadband network switch for the data case and to a PSTN switch in the telephony case. More specific details of typical air-protocols, the strategies service providers adopt to reduce end-to-end delays, and the changes to traffic characteristics when ATM cells are transmitted over the air will be covered in Chapter 3.

1.4   Need for Interworking

The trend toward ATM-based networks and the popularity of WANs have necessitated the specification and widespread implementation of interworking arrangements. Interworking schemes allow existing individual, small-business, and enterprise networks to continue to function and provide users the much-needed time-window to migrate to higher performance systems while making maximum use of the investment made in legacy systems.

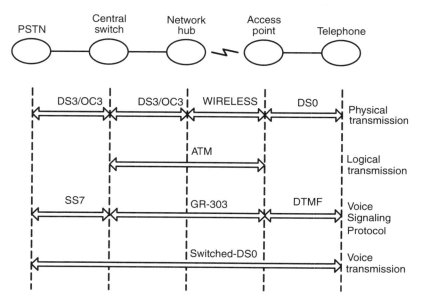

Figure 1.9 Wireless component in a typical voice network.

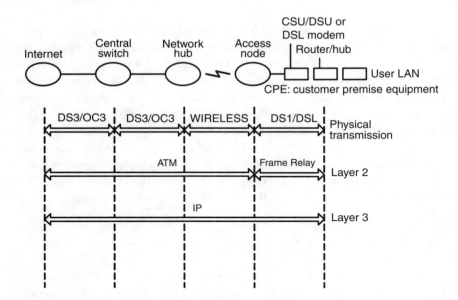

Figure 1.10 Wireless component in a typical data network.

Interworking needs can arise in different networking scenarios as identified in the following sections. Selected widely used interworking scenarios are given expanded treatment in Chapters 5–7.

1.4.1   Interworking Among Transport Data Networks

LANs in enterprises and small businesses play a dominant role in computer networking, and the interconnection of logical architectures of LANs to form different internetworking arrangements are increasingly becoming relevant in the marketplace.

Internetworking, in general, will encompass consideration of all seven layers of the OSI reference model. LANs comprise different types of bus, ring, and switched topologies, with Ethernet (IEEE 802.3) being one of the most popular bus-based technologies. Token Ring (IEEE 802.5) and Fiber Distributed Data Interface (FDDI; IEEE 802.6) are two other commonly available architectures that use a ring topology. The data-link layer (layer 2), which consists of sublayers [the hardware-independent logical link control (LLC) layer and the hardware-dependent media access control (MAC) layer] provides the necessary software abstractions to allow flexible interconnection and interworking schemes.

In bridged networks, LAN segments of the same or different types are interconnected with devices known as bridges. They include the following:

- Transparent bridges learn the MAC addresses of the different interconnecting LANs and use a bridging algorithm, such as the spanning tree, to identify a single route between stations and to avoid loops. The bridges themselves use bridge protocol data units (BPDUs) to communicate with other bridges to carry out the bridging algorithms.

- Source route bridging (SRB) is part of the IEEE 802.5 standard and was developed for token-ring networks.

- Translational bridging allows interconnectivity between unlike LAN segments.

- Source route transparent (SRT) bridging was developed to allow the SRT bridging technique to interoperate with translational bridges.

Routers are devices that generally operate at the network layer (layer 3) and have attained ubiquitous use in the widely popular IP networks to interconnect different subnets. Routers evolved to become the cornerstone of building WANs. With the proliferation of the Internet, routers increasingly added sophisticated functionality to interconnect different protocol nets. These include the following:

- Adaptation to encapsulate bridge protocols such as SNA and NetBIOS prior to transporting the data over IP, Frame Relay, or ATM-based transport nets;

- Interworking functionality between networking protocols such as X.25, IP, Frame Relay, and ATM.

Several standards have been written to facilitate interworking of network protocols. RFC 1577 provides the framework for point-to-point interworking of ATM-attached IP hosts in a single subnet environment. Next-Hop Resolution Protocol (NHRP) lays the framework for hosts in a multisubnet IP network to directly communicate via an ATM net.

Further, the above IP over ATM protocols are used in an Integrated Services Architecture (ISA) assisted by QoS-based protocols such as Resource Reservation Protocol (RSVP) to improve on the best-effort model of service delivery.

LANE is an effort by the ATM forum to facilitate interworking of legacy LANs with hosts via an ATM network. This specification is also restricted to a single LAN subnet.

In addition, the Frame Relay forum has defined Frame Relay to Frame Relay interworking (network interworking) via an ATM net as specification FRF.5, and also Frame Relay to ATM interworking (service interworking) as specification FRF.8. RFC 1490 defines encapsulation of IP over Frame Relay, and RFC 1483 defines encapsulation methods for IP over ATM. In addition, techniques to interwork SNA over Frame Relay and switched multimegabit data service (SMDS) over Frame Relay have also been defined.

Clearly, to enable continued use of existing networks while facilitating deployment of high-performance systems based on ATM, a bewildering array of interworking arrangements have been defined and are being put to increasing use.

1.4.2   Different Access Protocols

Besides internetworking of transport protocols, a related but not totally distinct segment where interworking will have a prominent role is the access networks of users at the edges of provider transport networks. When the user premise networks are of increased complexity, then the distinction between access segment interworking and network interworking will begin to blur.

Individual home users generally access service provider networks via dial-up accounts or, more recently, cable modems and different types of DSLs. Dial-up connections typically use TCP/IP running over PPP to the ISP, and the customer node uses the Dynamic Host Configuration Protocol (DHCP) to obtain the IP address the link will use for the established session. In these cases, any interworking or optimization of switch functions in the case of dial-up is performed at the service provider node and is transparent to the user.

Customers or businesses with increased bandwidth needs access service provider networks using IADs. These devices can integrate interworking functions to provide convenient multiplexing setup for different protocols so that the back end of the IAD provides a convenient single protocol interface to carrier networks, mainly ATM. The access devices can also be integrated with the wireless component to provide wireless access. Native ATM via T1, T3, or even OC-3 interfaces, Frame Relay on T1 or fractional T1 interfaces, and IP on 10/100 Mbps Ethernet interfaces are typical port access provisions available in the IADs. In this setup, Frame Relay may itself be carrying X.25 or IP, based on how the customer premises network evolved in the past. The back end to the carrier network will typically be ATM on OC-3c, IP over SONET, or IP over DWDM, either on wireless or fiber.

Circuit emulation service (CES) is an important type of service where interworking arrangements are used to transport structured T1/E1 with or without channel associated signaling (CAS), or unstructured T1/E1 (without framing information passing full T1/E1 bandwidth bit stream) transparently over ATM. In this scheme, voice carried in TDM slots can be transported and multiplexed through the ATM network with other packetized data. The key to this service, however, is whether the network can provide the QoS demanded by this application, which is termed as constant bit rate (CBR).

1.4.3   Voice Sessions and ATM

In addition to the direct TDM-based voice sessions to be carried over the ATM network using CES, the VoIP technology will increasingly use ATM networks due to the IP-ATM synergy.

In this section, we will briefly explore the possible interworking solutions available to enable IP telephony. Media gateway (MG), signaling gateway (SG) and media gateway control (MGC) form the three key components that facilitate PSTN-Internet interworking to allow IP telephony.

The lack of QoS guarantees in commercial IP nets will be a major hurdle for wide deployment of VoIP systems. In the interim, VoIP traffic will be increasingly carried by ATM nets, which can support the QoS needs of voice sessions. Challenges lie ahead in developing standards to facilitate development of ATM/PSTN gateways to support VoATM schemes. The architecture of the components assisting in the PSTN/data-net integration will likely remain the same as that developed for VoIP systems as shown in Figures 1.11(a, b)

Several standards that cover VoIP implementations exist. The family of H.323 standards was developed by ITU for use in broadband multimedia applications and was later adapted for voice. H.323 is the preferred protocol of traditional telephony providers. In contrast, the Session Initiation Protocol (SIP) developed by the IETF is lightweight and designed for easy integration with Hypertext Transfer Protocol (HTTP)–based applications. The Media Gateway Control Protocol, RFC 3015, is another IETF standard, while the ITU has developed the H.248 (MEGACO).

The media gateway function physically terminates the PSTN circuits on one end and acts as a host to the IP/ATM network at the other end. The softswitch (MGC), which has class 4 tandem switch capabilities and class 5

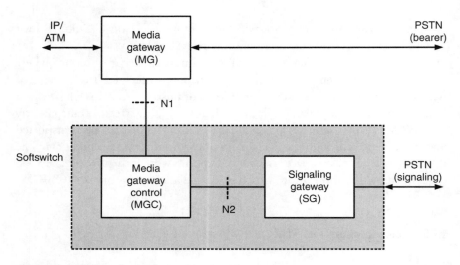

Figure 1.11(a) PSTN-data integration components.

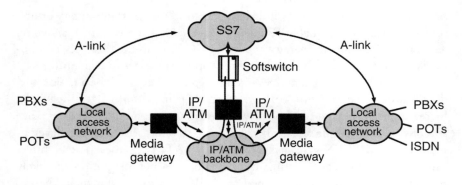

Figure 1.11(b) Media gateway and softswitch.

features, provides connectivity to the SS7 network and delivers the enhanced telephony services, such as 800-number lookup.

The potential explosion in the need to carry voice over ATM will be a continuing challenge to standards bodies to define workable internetworking standards during the next several years, as well as to the network service providers and router/switch/IAD vendors to implement cost-effective solutions for the consumer.

1.5 Traffic Management in ATM Networks

ATM is designed to be a multiservice network that can carry services with different QoS requirements. Unlike IP networks, including the Internet where only a best-effort philosophy is implemented, ATM attempts to achieve widely different QoS needs by techniques collectively called *traffic management*. Traffic-management techniques allow ATM networks to optimize bandwidth needs by judicious use of statistical multiplexing techniques and to enforce differing QoS needs of multiple application types.

At the connection level, traffic management uses different service classifications, QoS objectives, and traffic descriptors to derive a conformance definition for cells. The connection admission control (CAC) strategy is used to allocate the necessary bandwidth for a particular service.

At the cell level, the traffic is then policed to ensure that proper action is taken on the cells violating the conformance definition. In addition, the ATM layer can smooth the outgoing traffic to ensure that the conformance is established on cells going out of the network. Policing and smoothing functions are performed at the network edges at the user network interface (UNI) or network-to-network interface (NNI). With respect to a specific network, the traffic entering the network will be policed to limit the damaging impact on non-well behaved traffic sessions, and the traffic exiting the network will be smoothed so that outflowing traffic conforms to established parameters.

1.5.1 Impact of Interworking Protocol Characteristics

In protocol interworking scenarios where ATM is involved, traffic-management techniques employed at the non-ATM protocol segment will have an impact on the ATM end with its comprehensive traffic-management capabilities. The following factors need consideration:

- *Frame size.* Frame Relay or IP are variable frame/packet-oriented, and typically several ATM cells can be generated from a single frame/packet. For example, a 1,024-byte Frame Relay frame will result in about 20 ATM cells. Since this conversion likely takes place in hardware, there is a high likelihood of cells being produced in groups (known as clumping of cells), which violates ATM conformance rules.

- *Processing delays.* Frame Relay or IP frames require significant overheads in the layer 2 processing. In contrast to ATM networks, where switching is fully hardware-based, software overheads in Frame Relay and IP are likely to introduce large and varying interframe gaps that will result in undesirable cell traffic patterns.

- *Conversion overhead.* Adequate safeguards should be provided for the conversion of protocol traffic parameters to ATM parameters, the increase in ATM payload due to additional overhead introduced, and the fixed length of ATM cells.

These and other issues arising at the interworking nodes, when frames or packet flows are converted to cell traffic, will be critical to the QoS considerations of the end-to-end service and will be examined in detail in Chapters 8 and 9.

1.5.2   Impact of Wireless Segment

The modification of cell traffic patterns by the wireless component of the link will depend on the specific characteristics of the air-protocol employed to bridge the air-segment. Different aspects of the air-protocol will affect the traffic flow, impacting the end-to-end service. Some pertinent issues that need to be considered include the following:

- *Delay.* Air-protocol is likely to introduce delays, which will be unlike the processing delays in software layers. The delay will depend on the multiplexing technique used. If some form of FDM is used, there will be continuous transmission of data and the delay from the arrival of the frame/packet to the time at which the converted cells are transmitted over the air will be relatively minimal. However, if the TDM technique is used, the position of the assigned time slot's position when a cell arrives will determine the latency. Average delay will depend on the frame time, number of slots in a frame, and where the time slots required for the service are positioned. In addition to the initial delay due to the slot placement, the speed of the air link will introduce additional delays.

- *Clumping.* In addition to the clumping of cells due to the fast conversion of large frames into multiple ATM cells, clumping can be introduced by the format of air-protocol frames and the modulation techniques. Multiple cells can be carried in a specific TDM burst based on the scheme used. QUAD 64 QAM (four cells per burst in 64 QAM modulation) is an example. Slot assignments in the TDM frame will define the cell distribution at the terminating end of the air-segment.

Clumping can also be introduced due to multiplexing of different sessions through the same air pipe. If multiple sessions are established from a remote system and a hub, air-protocol definition may imply that the different connections use the same slot groups. If some of the connections do not use the maximum capacity of the defined session, other service connections may end up using the air slots that have become closely seated.

Chapters 2 and 3 provide a background to networking as it exists today and the coexistence of these networks with ATM-based broadband networks. Details of interworking techniques and the impact of wireless broadband component on the QoS issues of ATM networks form the core part of Chapters 5 through 9.

Selected Bibliography

Beninger, Toni, *SS7 Basics*, Chicago, IL: Intertec Publishing, September 1991.

The GSM Association, "GSM Is the Wireless Evolution," March 2002, at www.gsmworld.com/technology.

Sackett, George C., and Christopher Y. Metz, *ATM and Multiprotocol Networking*, New York: McGraw-Hill, 1996.

Spohn, Darren L., *Data Network Design*, New York: McGraw-Hill, 1993.

International Engineering Consortium, "Virtual Private Networks (VPNs)," April 2002, at www.iec.org/tutorials/vpn.

2

Background to Key Internetworking Protocols

2.1 Introduction

During its early stages of development, ATM appeared to have the potential to replace existing networking protocols as a new technology that would contain all required features to satisfy varying user needs. It was believed that ATM, in addition to its capability to support high-speed transport needs, would also extend itself into end systems, promoting the development of end applications that would use native ATM protocol features.

Instead, the evolution of network technology in the last several years, especially with the explosive growth in the use of the Internet, has led to a situation where IP remains the dominant technology for end users, and ATM has become the dominant transport network, carrying IP-datagrams efficiently between end systems.

ATM and IP have significant architectural differences that make interworking arrangements challenging. ATM is a connection-oriented protocol, whereas IP is connectionless. There are also differences in addressing and resource allocation. Several different standards, which are briefly summarized below, have been defined to allow ATM-IP interworking:

- Simple encapsulation of IP in ATM protocol (RFC 1483 [1]). ATMARP for mapping ATM and IP addresses was added in RFC 1577 [2]. These RFCs were limited to a single IP-subnetwork.

- Interconnecting multiple IP-subnetworks with intervening ATM required a next-hop router to travel between the two subnets. The NHRP [3] eliminated the need for a next-hop router.

- LANE [4, 5] standards were developed to enable ATM networks to appear as a multiaccess LAN.

- Generic ATM interworking with other network layer protocols gave rise to multiprotocol over ATM (MPOA) standards [6].

- Integrated services (IntServ) and differentiated services (DiffServ) [7] models enforce different QoS provisions for IP networks. The ATM architecture provides for different QoS service categories. A new protocol, RSVP [8], was specified to facilitate QoS capability for IP.

The complexity of mapping between IP and ATM networks is clear from these different attempts to define procedures for specific configurations of networks. Multiprotocol label switching (MPLS) [9] is a technique that simplifies the mapping by allowing ATM switches to switch IP-datagrams at the link level, bypassing the need for ATM-signaling protocols. The technique is based on IP-switching promoted by Ipsilon and tag-switching advanced by Cisco Systems. MPLS involves switching at link layers within the core ATM transport network. MPLS facilitates ATM-IP interworking at performance levels far better than those provided by IP-routers and avoids signaling protocol mapping between ATM and IP.

IPSec [10] is another protocol extension that has gained importance in the IP world. IPSec defines two protocol extensions—Authentication Header (AH) and Encapsulating Security Payload (ESP)—that provide security to entire IP-payload (tunnel mode) or to transport IP-payload (transport mode).

This chapter will outline IP, Frame Relay and ATM protocols that have become relevant to the interworking scenarios present in current networks. In particular, the details of IPv4 (the current version of the IP protocol), the enhanced next-generation IPv6 protocol, and their main differences will be outlined. The supporting protocols that allow routing within an autonomous region, defined by Interior Gateway Protocols (IGP) such as RIP and OSPF, will be discussed. The Border Gateway Protocol (BGP), an interdomain routing protocol that passes information between two autonomous networks, will also be discussed.

Basics of the ATM protocol will also be described, along with a brief outline of the evolving signaling protocols within ATM. ATM addressing and associated routing protocols are not yet widely deployed and will only be briefly noted.

Frame Relay has long served as an access and transport protocol before ATM gained popularity. Its simplicity and efficiency in carrying data in a reliable physical transport medium and its wide penetration into the customer premise dictate the need to carry Frame Relay data over ATM transport networks. Therefore, Frame Relay also will be outlined in this chapter.

2.2   Introduction to IP

This section presents an overview of the connectionless IP suite. We start by describing the early origins of the IP, followed by the packet formats of IPv4 and IPv6. The differences in the protocol features of IPv4 and IPv6 are described. In this section, both *internet* and *Internet* refer to networks that are interconnected by TCP/IP.

2.2.1   IP Overview

TCP/IP has become the dominant end-system protocol stack, due to the Internet's rapid growth and open protocol standards that were developed independently from any specific computer hardware or operating system. TCP/IP's independence from any specific physical network hardware allows it to run over Ethernet, a dial-up line, X.25 and Frame Relay networks, ATM, Token Ring, and virtually any other kind of physical transmission media.

Figure 2.1 illustrates the various layers in the Internet protocol suite built for IP. The User Datagram Protocol (UDP), Internet Control Message Protocol (ICMP), routing control protocols, and Transmission Control Protocol (TCP) interface directly with IP, compromising the transport layer in the Internet architecture. A number of applications interface to TCP, as shown in Figure 2.1. The File Transfer Protocol (FTP) application provides for client-server login, directory listing, and file transfers. Telnet provides a remote terminal login capability, similar to the old command line interface of terminals to a mainframe. HTTP supports the popular WWW. These three protocols operate over TCP, a reliable transport protocol that provides

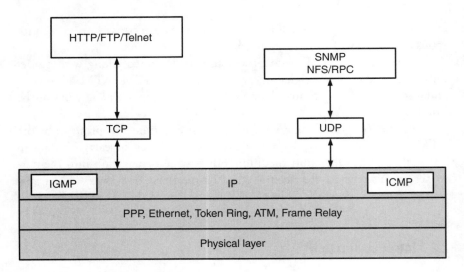

Figure 2.1 Mapping of the IP protocol.

a connection with sequencing, retransmission, and error protection (checksums).

The UDP, which also runs over IP, only provides an unacknowledged datagram capability. Application data sent using UDP has no retransmission capabilities when there is loss due to errors or congestion in the network. The Simple Network Management Protocol (SNMP), Remote Procedure Call (RPC), and Network File Server (NFS) are examples of Internet application protocols that run over UDP.

The ICMP is used by routers to send error and control messages to other routers. ICMP also provides a function in which a user can send a ping (echo packet) to verify host reachability and round-trip delay. The IP layer also supports the Internet Group Multicast Protocol (IGMP).

In summary, IP is the backbone protocol for higher layer protocols such as TCP and UDP, which form the main IP suite. TCP provides reliable transmission over IP compared to UDP, which simply provides a datagram service over IP.

The version of IP that is most commonly used today is IPv4, also known as classical IP. Figure 2.2 illustrates the format of the IPv4 packet. The IP header is five or six 32-bit words long (the sixth word is optional). The header contains all the necessary information to deliver the packet. The following lists specific details about IPv4:

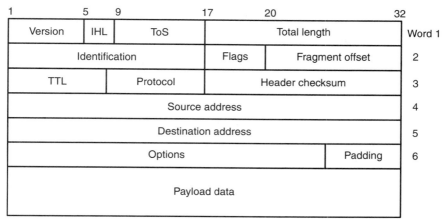

IHL = IP header length

Figure 2.2 IPv4 format.

- *Type-of-service (ToS) field.* The standard IPv4 contains a 3-bit precedence field, plus three separate bits specifying other services attributes and two unused bits. The precedence field ranges from 0 (i.e., normal priority) to 7 (i.e., network control), indicating eight levels of precedence. The three individual bits request low delay, high throughput, and high reliability. Six bits of the ToS field are redefined as the DiffServ code point (DSCP) in IP DiffServ [7], which provides a means for IP routers and hosts to differentiate among various classes of IP traffic in order to control QoS characteristics such as latency, bandwidth, and packet discard strategies. For example, IP DiffServ [7] policy provides preferred treatment to VoIP traffic.

- *Total-length field.* This specifies the total IP datagram length for the header plus the user data.

- *Identification field, flags, and fragment-offset fields.* These control fragmentation (or segmentation) and reassembly of IP datagrams.

- *Time-to-live (TTL) field.* This specifies how many hops the packet can be forwarded in the network before declaring the packet dead and, hence, discard-eligible. Typically, intermediate nodes decrement the TTL field at each hop. When the TTL field reaches zero, intermediate nodes discard the packet.

- *Protocol field.* This identifies the higher-level protocol (e.g., TCP or UDP) that specifies the format of the payload data.

- *Header checksum.* This ensures the integrity of the header fields through a simple bit-parity check.

The IP delivers the datagram by checking the destination address in word five of the header. The destination address is a standard 32-bit IPv4 address that identifies the destination network and the specific host on that network.

The Internet was running out of network numbers with IPv4, the routing tables were getting too large, and there was a risk of running out of addresses altogether. Its successor, IPv6, is not a simple derivative of IPv4, but a definitive improvement. For example, a fixed format is assigned to all headers in IPv6 and does not contain any optional element. Options for special-case packets are specified using extension headers (appended after the main header) unlike in IPv4, which uses a variable length optional field. As a result, there is no need in IPv6 for a header length field (IHL). The header checksum is removed in IPv6 to diminish the cost of header processing, as there is no need to check and update the checksum at each switching node. Checksums in the media access control procedures of IEEE-802 [11] networks, in the adaptation layers of ATM, and in the framing procedures of the PPP for serial links provide adequate lower-layer error detection.

IPv6 supports all the traditional protocols that IPv4 did, such as UDP datagram service and TCP services like FTP file transfers, e-mail, X-windows, Gopher, and the Web.

IPv6 expands the address field size from 32 to 128 bits, described in Section 2.2.2. The addition of a scope field allows easier multicast routing. IPv6 supports the anycast feature, where a host sends a packet to an anycast address, which the network delivers to the closest node supporting that function. IPv6 has the capability to define QoS for individual traffic flows using RSVP [8].

Figure 2.3 illustrates the format of the IPv6 packet. The new header is much simpler than that of the classical IPv4. The only field that keeps the same meaning is the version number, which, in both cases, is encoded in the very first four bits of the header. IPv4 fields (i.e., the header length, type of service, identification, flags, fragment offset, and header checksum) are suppressed in the IPv6 header. Three IPv4 fields—the length, the protocol type, and the TTL—are renamed and redefined. The payload length in the IPv6 header indicates the number of bytes following the required 40-byte header.

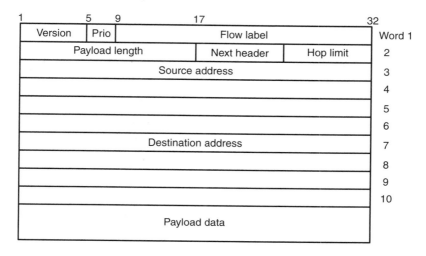

Figure 2.3 IPv6 format.

The next header field identifies the subsequent header extension field. There are six (optional) header extensions: hop-by-hop options, source routing, fragmentation support, destination options, authentication, and security support. The last extension field identifies the higher-layer protocol type using the same values as IPv4, typically TCP or UDP. The hop limit field determines the maximum number of nodes a packet may traverse. Nodes decrement by 1 in the hop limit field each time they forward an IP packet, analogous to the TTL field in IPv4, discarding the packet if the value ever reaches zero.

The option mechanism is entirely revised, and two new fields, a priority and a flow label, have been added to facilitate the handling of real-time traffic. The priority field has 16 possible values. By and large, it plays the same role as the precedence field of IPv4. The flow label is used to distinguish packets that require the same treatment; that is, they are sent by a given source to a given destination with a given set of options. The 4-bit priority field defines eight values (0–7) for sources that can be flow controlled during congestion, and eight values (8–15) for sources that cannot respond to congestion (e.g., real-time constant bit-rate traffic, such as video). Within each priority group, lower numbered packets are less important than higher numbered ones. The flow label is intended for use by protocols such as the RSVP [8] that provide for QoS over IP packets involved in the same flow.

In summary, changing the IP from IPv4 to IPv6 has some impact on upper layers. This impact is minimal, because the IPv6 datagram service is identical to the classical IPv4 service. However, the implementation of transport protocols such as TCP or UDP will have to be updated to take into account the larger addresses. Finally, IPv6 is a more efficient protocol than its predecessor IPv4. Going from 32 to 128 bits for addressing will make it possible to number thousands of billions of hosts, as well as provide room to insert many more degrees of hierarchy than the basic three layers (network, subnet, and host) accommodated by IPv4.

2.2.2   IP Addressing

The number of address bits used to identify the network, and the number used to identify the host, vary according to the class of the address. The main address classes are class A, class B, and class C (see Table 2.1). By examining the first few bits of an address, the IP routing software can determine the address's class and, therefore, its structure. The class determines the maximum network size, measured by the number of hosts.

The central authority, the Internet Assigned Numbers Authority (IANA), assigns IP addresses (IPv4) to ensure that public IP addresses are unique. A network administrator then assigns the host address within its class A, B, or C address space however he or she wishes, as long as the assignment is unique.

This inefficiency of arbitrary IP address assignments led to large blocks of unused addresses. Furthermore, many organizations exacerbated the problem by requesting valuable class A and C addresses and then inefficiently assigning them, leaving large block of addresses unused. Due to an ever-

Table 2.1
IP Address Classes

Address Class	First Byte Value (Decimal)	Network Address	Host Address
A	1–127	Byte 1 127 Class A networks	Bytes 2–4 16,777,214 hosts each
B	128–191	Bytes 1–2 65,000 Class B networks	Bytes 3–4 65,534 hosts each
C	192–223	Bytes 1–3 2,000,000 Class C networks	Byte 4 254 hosts each

increasing demand and the diminishing of address space, these unused class A and C addresses were reused by new service providers by dynamically allocating addresses to individual users using PPP. This new classless addressing was called classless interdomain routing (CIDR).

Instead of being limited to network identifiers of 8, 16, or 24 bits, CIDR currently uses prefixes anywhere from 13 to 27 bits. Thus, blocks of addresses can be assigned to networks as small as 32 hosts or to those with more than 500,000 hosts. This allows for address assignments that much more closely fit an organization's specific needs. A CIDR address includes the standard 32-bit IP address, as well as information on how many bits are used for the network prefix. For example, in the CIDR address 206.13.01.48/25, the /25 indicates the first 25 bits are used to identify the unique network, leaving the remaining bits to identify the specific host (see Table 2.2).

The IANA has reserved the following three blocks of the IP address space for private internets for hosts that do not require access to hosts in other enterprises or the Internet:

10.0.0.0–10.255.255.255

172.16.0.0–172.31.255.255

192.168.0.0–92.168.255.255

The first block is referred to as *24-bit block,* the second as *20-bit block,* and the third as *16-bit block*. Addresses within this private address space will only be unique within the enterprise (RFC 1918 [12]). The advantage of using private address space for the Internet at large is to conserve the globally unique address space by not using it when it is not required. As before, any enterprise that needs a globally unique address is required to obtain one from an Internet registry. An enterprise that requests IP addresses for its external connectivity will never be assigned addresses from the blocks reserved for private internets.

Since IPv6 is based on the same architecture principle as the classic IPv4 protocol, one could very well expect IPv6 addresses to be larger versions of IPv4 addresses. Like IPv4 addresses, IPv6 addresses identify an interface connected to a subnetwork, not a station. As in IPv4, a station that is multi-homed will have as many addresses as interfaces. One big difference is that IPv6 routinely allows each interface to be identified by several addresses to facilitate routing or management. IPv6 addresses belong to one of three categories:

Table 2.2
Classless Interdomain Routing

CIDR Block Prefix	Number of Equivalent Class Cs	Number of Host Addresses
/27	$\frac{1}{8}$	32
/26	$\frac{1}{4}$	64
/25	$\frac{1}{2}$	128
/24	1	256
/23	2	512
/22	4	1,024
/21	8	2,048
/20	16	4,096
/19	32	8,192
/18	64	16,384
/17	128	32,768
/16	256	65,536
/15	512	131,072
/14	1,024	262,144
/13	2,048	524,288

- *Unicast*—point-to-point addresses identifying exactly one interface.
- *Multicast*—a group of stations. A packet sent toward a multicast address will normally be delivered to all members of the group. Multicasting capabilities were added to IPv4, using class D addresses and the IGMP. IPv6 defines a multicast address format that all routers should recognize and incorporates all functions of IPv4's IGMP into the basic ICMP of IPv6.
- *Anycast*—a group of stations. The difference between multicast and anycast is in the transmission process. Instead of being delivered to all members of the group, packets sent to a unicast address are

normally delivered to only one point, the nearest member of the group. This new feature offers a lot of flexibility to network managers. One could use anycasting to find out the nearest DNS server, the nearest file server, or the nearest time server. There is no specific anycast format in IPv6. Anycast addresses are treated exactly the same way as unicast addresses. The new responsibilities are on the router, which has to maintain one route for each anycast address that is active in a given subnet.

2.2.3   IP Routing

Routing is the glue that binds the Internet together. Without it, TCP/IP traffic would be limited to a single physical network. Routing allows traffic from your local network to reach its destination somewhere else in the world after passing through many intermediate networks. The important role of routing and the complex interconnection of Internet networks make the design of routing protocols a major challenge to network software developers. Consequently, most discussions of routing concern protocol design. Very little is written about the important task of properly configuring routing protocols. However, more daily network problems are caused by improperly configured routers than by improperly designed routing algorithms. This section describes different types of routing protocols that have evolved for IP-based networks. A distinction should be made between routing and routing protocols. All systems route data, but not all systems run routing protocols. Routing is the act of forwarding datagrams based on the information contained in the routing table. Routing protocols are programs that exchange the information used to build routing tables.

The building block of the IPv4 routing as shown in Figure 2.4 is the autonomous system (AS), a collection of subnetworks managed by a single entity. An autonomous system may be, for example, the network of an ISP or that of one large company. Routing protocols are divided into two general groups: interior (intradomain) and exterior (interdomain) protocols.

An interior protocol is a routing protocol used within an autonomous system (domain). The RIP is a widely used interior routing protocol. It is well suited for LANs. RIP selects the route with the lowest hop count as the best route based on a distance vector algorithm where neighbor nodes periodically exchange vectors of the distance to every destination subnetwork. This process eventually converges on the optimal (shortest) routes. The RIP hop count represents the number of nodes through which data must pass to

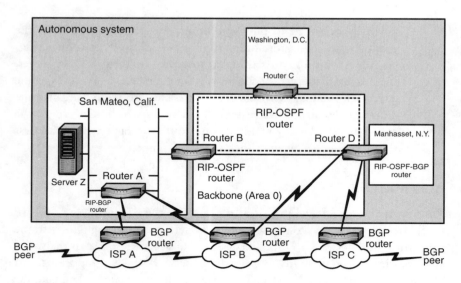

Figure 2.4 AS and interior/exterior routing protocols. (Reprinted with permission of Network Computing © 2002.)

reach its destination. RIP assumes that the best route is the one that uses the fewest nodes (i.e., fewer nodes mean a shorter path), and the shortest path is the best path. The longest path RIP accepts is 15 hops. If the route is greater than 15 hops, RIP considers the destination unreachable and discards the route. For this reason, RIP is not suited for very large autonomous networks.

A second method is where each router learns the entire link state topology of the entire network. Unlike RIP, this method is well suited for very large networks. The OSPF protocol is an example of a link state protocol that provides equal cost multipath routing. OSPF can maintain several routes to the same destination. It reliably floods an advertisement throughout the network whenever a state of a link changes (e.g., adding a new link, deleting a link, or having an unexpected link failure). Thus, each OSPF router obtains complete knowledge of the network topology after a certain convergence time, which is usually in seconds. After any change, each router computes the least-cost routes to every destination using an algorithm such as the Dijkstra algorithm. Since every router has the same topology database, they all compute consistent next-hop forwarding tables.

The RIP for IPv6 is a straightforward evolution of IPv4 RIP. The simple distance vector technology of RIP is inferior to the link state technology of OSPF. As a result, OSPF became the recommended routing protocol for

IPv6. The IPv6 version of OSPF is a simple translation of IPv4 OSPF, making minimal changes to accommodate the new IPv6 address format. OSPF for IPv6 runs on top of IPv6, between IPv6-capable nodes. The IPv6 link state database is not shared with the IPv4 database.

Exterior routing protocols, also known as interdomain routing protocols, are used to exchange routing information between autonomous systems. RFC 1267 [13] defines the BGP, an exterior routing protocol that exchanges reachability information between autonomous systems. BGP provides information about each route; routers can use these path attributes to select the best route. Policy-based routing uses nontechnical attributes and criteria such as organizational and security considerations to make routing decisions. BGP is optimized to handle IPv4 32-bit addresses and could not be easily upgraded to handle IPv6. For this reason, the exterior protocol for IPv6 is based on the Interdomain Routing Protocol (IDRP), which was first designed as a component of the OSI family of protocols.

2.2.4 Link-Layer Protocols for IP

As previously described, the Internet protocol can be run over various link-layer protocols such as Ethernet, a dial-up line, serial links, X.25 and Frame Relay networks, ATM, Token Ring, and virtually any other kind of physical transmission media. The IP datagrams are encapsulated in the relevant data-link layer protocol header and simply shipped to the link endpoint where datagrams are decapsulated.

The PPP not only supports IP, but other network protocols multiplexed over a shared serial link. PPP supports automatic configuration and management of the link layer between dial-up users, as well as between multiprotocol routers connected by a wide range of serial interfaces. PPP allows an ISP to efficiently share a limited IP address space, since only users who are logged in actually use an IP address. Therefore, an ISP only needs the number of IP addresses equal to the number of dial-up ports in use and not for the total number of subscribers. This feature of PPP stretches the limited address space of IPv4. The authentication feature of PPP allows the network to check and confirm the identity of users attempting to establish a connection. Today, PPP is ubiquitous; Web browsers, operating systems, hubs, routers, and servers have embraced PPP as the link layer of choice for accessing the Internet.

The DHCP [14] provides the means for passing configuration information to a client, such as a computer on a TCP/IP network. A DHCP client is a program that is located in (downloaded to) each computer so that it can be configured. One of the main features of DHCP is to provide the capability of automatic allocation of reusable network addresses instead of being given a static IP address. When the client boots up, it sends a BootP [15] request for an IP address. A DHCP server then offers an IP address that has not been assigned from its database, which is then leased to the client for a predefined time period. DHCP supports three mechanisms for IP address allocation. In *automatic allocation,* DHCP assigns a permanent IP address to a client such as a computer containing a Web server that needs a permanent IP address. In *dynamic allocation,* DHCP assigns an IP address to a client for a limited period of time (or until the host explicitly relinquishes the address). Dynamic allocation is particularly useful for assigning an address to a client that will be connected to the network only temporarily or for sharing a limited pool of IP addresses. In *manual allocation,* a client's IP address is assigned by the network administrator, and DHCP is used simply to convey the assigned address to the host. A particular network will use one or more of these mechanisms, depending on the policies of the network administrator. The available IP addresses are stored in a central database (DHCP server), along with information such as the subnet mask, gateways, and DNS servers.

2.3   Introduction to Frame Relay Protocol

This section presents an overview of Frame Relay, a popular low-cost packet switching protocol for connecting devices on a WAN. Frame Relay has taken much of the public packet-switching communications market away from X.25. As we will see in this section, the Frame Relay Protocol has reduced the complexity of the link-layer protocol overhead to be more efficient than X.25. Public Frame Relay services have displaced many private line-based data networks, thus giving profound importance for Frame Relay in data networks.

Developed in the early 1990s, Frame Relay filled a technology gap between X.25 and ATM. Frame Relay interworking with ATM as described in Chapter 6 facilitates the smooth transition from X.25 to ATM. Support for both X.25 and Frame Relay allows Frame Relay to be used where noise quality is too high for X.25 to be used. Frame Relay provides greater granularity for bandwidth allocation, compared to X.25 or another predecessor, SNA.

Frame Relay networks were developed to address the growing need for increased speed and to leverage the availability of very clean transmission lines. The demand for higher speed data communications is driven by the move from textual to graphics interactions, the increase in bursty traffic applications, and the proliferation of LANs and client-server computing. Bursty traffic allows statistical multiplexing of bandwidth. By eliminating error correction procedures on every link in the network, the amount of processing at every node can be minimized.

Frame Relay's enormous success is primarily due to its use as a replacement for leased lines for interconnecting local-area data networks. Frame Relay has an established track record of carrying IP traffic. It is also used for carrying legacy protocols, such as async, bisync, and high-level data-link control (HDLC). IBM's SNA protocols, which are frequently used for financial and business applications, are also being carried by Frame Relay.

Frame Relay is not as well suited as ATM for carrying traffic types such as voice and digital video that are intolerant of variable delay. Frame Relay provides an upgrade to existing X.25 and private-line packet-switch technology by supporting speeds ranging from 64 Kbps through nxDS1/nxE1 and all the way up to DS3/E3. Frame Relay is a better choice to link fairly widely separated locations, because its pricing structure is usually insensitive to distance; that is, it does not have the mileage rates usually attached to the private line tariffs.

Frame Relay network consists of endpoints (e.g., PCs, servers, host computers), Frame Relay access equipment (e.g., bridges, edge routers, Frame Relay access devices), and network devices (e.g., switches, network routers, T1/E1 multiplexers). A Frame Relay network is often depicted as a network cloud, as it is not a single physical connection between one endpoint and the other. Instead, a logical path is defined through the network. This logical path is called a virtual circuit (VC). Bandwidth is allocated to the VC until actual data needs to be transmitted as described in the Frame Relay traffic-management section. Frame Relay is well suited to carry TCP/IP, X.25, and SNA traffic over its networks, as shown in Figure 2.5.

2.3.1 Frame Relay Protocol Overview

Frame Relay is a link-layer protocol that has no sequencing and no retransmission to detect and recover from misordered or lost frames. Higher layer protocols, like TCP, must perform retransmission and sequencing. Frame Relay virtual circuits—both PVCs and SVCs—are the basis for identifying logical connectivity between users. PVCs and SVCs are supported by Frame Relay data-link protocols. Finally, the traffic-management aspects

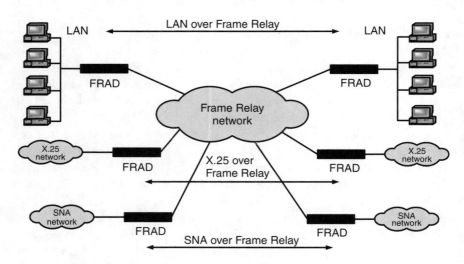

Figure 2.5 Frame Relay network.

describing Frame Relay rate enforcement mechanisms are presented in this chapter.

2.3.2   Frame Relay PVCs

PVCs in Frame Relay are preconfigured by a network operator. Provisioning PVCs requires thorough planning, a knowledge of traffic patterns, and bandwidth utilization. Once the PVC is provisioned, data can be sent by the two endpoints using Frame Relay Q.922 [16] procedures. A data-link connection identifier (DLCI) identifies the bidirectional Frame Relay connection at the interface between user device and the Frame Relay network. In the example shown in Figure 2.6, user A communicates with user B using DLCI 40, while user B communicates with user A using DLCI 55. When a frame is ready to be transmitted from user A to user B, the user A device assigns the value 40 to the DLCI in the frame. When received by the network, the frame is sent to the network's egress point, where the DLCI is changed to 55 before being forwarded to the destination user device.

DLCIs have only local significance and must be unique for all connections between the same endpoints. Figure 2.6 illustrates this property with two other connections. User A must use a different DLCI (e.g., DLCI 41) to communicate with user C. However, user C may communicate with user D using DLCI 40.

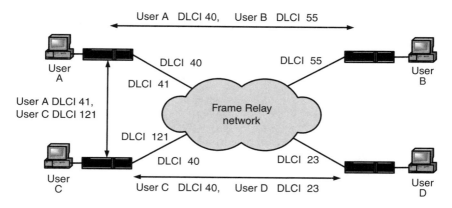

Figure 2.6 Frame Relay DLCI connection identifiers.

2.3.3 Frame Relay SVCs

Applications that do not require dedicated bandwidth capabilities provided by PVCs can establish dynamic SVCs. SVCs do not require preconfiguration; they use Q.933 [17] signaling to establish temporary point-to-point connections to any addressable node connected to the Frame Relay network. SVCs require the use of the E.164 (ISDN/Telephony Numbering Plan) or the X.121 (Data Numbering Plan) addresses.

An SVC is initiated when an endpoint sends a setup message to the Frame Relay network. Setup includes the destination address, bandwidth parameters, and other information about the call and its use. The network sends a call-proceeding message back to the calling user, along with the DLCI assigned to the connection. If the receiving party accepts the call, a connect message is sent from the receiving user back through the network to the calling user, indicating acceptance. Once the connection is set up, the two endpoints can transfer information. When both parties have completed the call, either endpoint can send a disconnect message to the network. The network releases the resources of the call and sends a release message to both endpoints. When the network receives a release-complete message from both endpoints, the call is terminated and the connection no longer exists.

2.3.4 Protocols for Frame Relay

Frame Relay is efficient because it eliminates all OSI layer 3 processing. Only a few layer 2 functions, the so-called core aspects, are used, such as checking for a valid, error-free frame, but not requesting retransmission if an error is found. Thus, many protocol functions already performed at higher levels, such as sequencing, window rotation, acknowledgments, and supervisory frames, are not duplicated within the Frame Relay network.

The protocol data units in Frame Relay are called *frames* and are described below:

1. A frame consists of an opening flag followed by an address field, a user data field, an FCS, and a closing flag as shown in Figures 2.7(a–c).

2. The address field consists of two, three, or four octets.

   - The DLCI is used to identify the VC for the Frame Relay connection.

   - The command/response (C/R) indication bit may be set to any value by the higher layer and is carried transparently by the Frame Relay network.

   - The discard-eligibility (DE) bit is set to 1 to indicate that a frame should be discarded before other frames in a congestion situation (i.e., when frames must be discarded to ensure safe network operation and maintain the committed level of service within the network).

   - The backward explicit congestion notification (BECN) bit is set to 1 by a congested network to notify the user that congestion avoidance procedures should be initiated for traffic in the opposite direction of the transmitted frame.

   - The forward explicit congestion notification (FECN) bit is used by a congested network to notify the user that congestion avoidance procedures should be initiated where applicable for traffic in the direction of the transmitted frame. This bit may be used with destination controlled transmitter rate adjustment.

   - The extended-address (EA) bits are used to delimit the address field. For example, the EA bit in the first octet is set to 0 and the EA bit in the second octet is set to 1 in the two-octet address format.

3. The user data field consists of an integral number of octets. This payload is not interpreted by the Frame Relay data-link layer.

4. The FCS is a two-octet field. The FCS covers all of the bits in the frame existing between, but not including, the final bit of the opening flag and the first bit of the FCS.

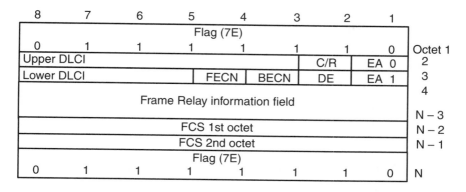

Figure 2.7(a) Frame Relay format with two-octet address.

Figure 2.7(b) Frame Relay format with three-octet address.

Figure 2.7(c) Frame Relay format with four-octet address.

2.3.5 Local Management Interface

The local management interface (LMI) operates between the network and the user devices. The purpose of LMI is to provide the user with status and configuration information concerning the DLCIs available at this interface. The messages transferred via the LMI protocol provide the following features:

1. Notification of the network to the user device of active and inactive DLCIs;

2. Notification of the user device of the removal or addition of a DLCI;

3. Link integrity verification (LIV) procedures that provide real-time monitoring of the status of the physical and logical link between the network and each user device by exchanging a sequence-numbered keep-alive between the network and the user device.

The LMI described above only provides for one-way querying and response, meaning that only the user device [or customer premise equipment (CPE)] can send a status-inquiry message. The ITU LMI only addresses the UNI and would not work in a NNI due to this one-way communication. UNI provides the end-device interface to the network. NNI provides the ability for networks to query and respond to one another. ANSI extended LMI to provide a bidirectional (symmetric) mechanism for PVC status signaling, which allows both sides of the interface to issue the same queries and responses.

There are three versions of the LMI specifications:

1. American National Standards Institute (ANSI) T1.617 Annex D [18];

2. ITU-Telecommunications Services Sector (ITU-T) Q.933 Annex A [17];

3. The LMI as defined in FRF.1 superseded by FRF.1.1 [19] (the revised FRF.1.1 [19] calls for the mandatory implementation of Annex A of ITU Q.933).

Each version defines a slightly different use of the management protocol. To ensure interoperability, the same version of management protocol must be at each end of the Frame Relay link. In a UNI configuration, one-way querying is used, and in a NNI configuration, the bidirectional querying is used. The data-link layer of the LMI conforms to Q.922 [16] data-link layer for Frame Relay. Frame Relay DLCI zero is used for the LMI exchanges.

2.3.6   Traffic Management in Frame Relay

ANSI Recommendation I.370 [20] defines traffic-management procedures (rate enforcement) for Frame Relay traffic. Traffic that enters the network for each Frame Relay PVC and SVC is monitored against traffic parameters configured for the PVC or signaled for the SVC. These include

- *Committed information rate (CIR):* The maximum average rate (in bits per second) at which the network promises to this end of the connection to send data for delivery;
- *Committed burst size (Bc):* The maximum amount of data (in bits) that the network agrees to deliver over a period *T* seconds for this end of the connection, where *T* = Bc/CIR;
- *Excess burst size (Be):* The maximum amount of data (bits) in excess of Bc that the network will attempt to deliver over a period *T* for this end of the connection, where *T* = Bc/CIR.

As shown in Figure 2.8, the Frame Relay backbone attempts to carry (but does not guarantee) traffic delivered within the configured CIR as long as individual bursts do not exceed the configured Bc value. Frames in excess of Bc, up to Be, are marked DE and are discarded in preference to unmarked frames when there is congestion within the backbone. Frames in excess of Be are discarded locally before entry to the network. Frames that enter the switch with the DE bit already set are counted directly against Be.

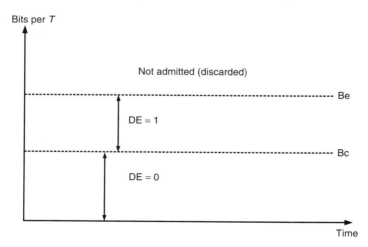

Figure 2.8 Frame Relay traffic management.

The intent of the traffic parameters is to prevent one Frame Relay connection from using excessive network resources at the expense of another connection. This standardized mechanism makes it more likely that the connections that deliver traffic within their configured Bc do not lose frames due to network congestion. Connections can exceed their configured Bc as long as there is available network capacity. However, traffic management ensures that frames with DE marked cannot crowd out frames that do not have DE marked.

2.3.7 Frame Relay Standards and Interworking Agreements

The Frame Relay Forum develops and approves implementation agreements (IAs) to ensure Frame Relay interoperability and facilitates the development of standard conformance tests for various ITU and ANSI Frame Relay protocols. Since the earliest IAs, additional features, such as multicast, multiprotocol encapsulation, and switched virtual signaling, have been defined in subsequent IAs to increase the capabilities of Frame Relay.

Figure 2.9 shows Frame Relay standards that are applicable to a typical Frame Relay network configuration.

The following briefly describes interface interworking and encapsulation standards that the Frame Relay Forum developed over the years for the implementation of Frame Relay:

1. *FRF.1.2:* The PVC UNI agreement defines the interface between Frame Relay devices; for example, one at the customer premises and one at the central office. The UNI protocol allows users to access a private or a public Frame Relay network.

2. *FRF.2.1:* The PVC NNI agreement defines how two different Frame Relay networks should communicate.

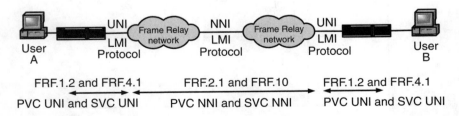

Figure 2.9 Frame Relay UNI and NNI configuration standards.

3. *RFC 1490:* The IETF developed a standardized method to encapsulate various protocols in Frame Relay.

4. *FRF.3.1:* ANSI and the Frame Relay Forum enhanced RFC 1490 to include support of the SNA protocols as defined in Frame Relay multiprotocol encapsulation. FRF.3.1 is used to carry SNA traffic across a Frame Relay network and may also be used to transport IP.

5. *FRF.4.1:* The SVC UNI agreement defines the mechanisms by which user devices can request dynamically established virtual circuits to be set up across a Frame Relay network.

6. *FRF.5:* Network interworking allows connection of two Frame Relay endpoints such as Frame Relay access devices (FRADs) or routers, which are attached to a Frame Relay network, over an ATM backbone. The FRADs have no knowledge of the ATM backbone because the network equipment, particularly ATM WAN switches, provides the interworking function. Frame Relay/ATM network interworking is defined jointly by the Frame Relay and ATM Forums as described in Chapter 6.

7. *FRF.8:* Service interworking connects a Frame Relay network to an ATM network, allowing Frame Relay devices to communicate with ATM devices. With service interworking, users can coexist with or migrate a portion of the existing Frame Relay network to ATM without requiring any special software in either end-user device. See Chapter 6 for details on service interworking.

8. *FRF.9:* Defines the encapsulation of the Data Compression Protocol (DCP) over Frame Relay.

9. *FRF.10:* The SVC NNI defines SVC NNI procedures to be used between Frame Relay networks.

10. *FRF.11:* VoFR offers a consolidation of voice and voice-band data (e.g., fax and analog modem) with digital data services over the Frame Relay network. FRF.11 establishes the standard for Frame Relay voice transport.

2.4 ATM

ATM is a technology that offers the bandwidth-on-demand features of packet-switching with the high speeds required for today's LAN and WAN networks. This cell-relay technology operates independently of the type of transmission being generated at the upper level and the type and speed of the physical layer medium being used. This allows sending of virtually any type of transmission (e.g., voice, data, video) in a single integrated data stream operating over any medium ranging from T1/E1 lines to high-speed optical signals such as SONET at OC-3, OC-12, OC-48, OC-192 and OC-768 rates. ATM technology permits both public and private networks to provide seamless, transparent connections from one end user to another, whether endpoints are located in the same building or across two cities.

ATM is considered to be mature today. It provides a stable and well-tested implementation of QoS and traffic management that serves integrated networks. ATM is widely used to carry other data services like Frame Relay and provides an excellent network access technology due to its ability to integrate voice, data, and video over the same network at different speeds, guaranteeing QoS requirements. ATM allows user data to consume bandwidth only when needed. Unused bandwidth is consumed by other applications by applying statistical multiplexing capabilities. ATM is asynchronous because each cell can be independently addressed to allocate bandwidth across virtual channels as needed. Contrast this to FR PVCs that statistically allocate bandwidth. ATM traffic is conveyed in 53-byte packets known as cells. ATM uses fixed-length cells for transmission to enable switching in hardware at a much higher rate than conventional software switching. Small fixed-length cells enable networks to have predictable response times, making ATM ideal for such applications as digitized voice and video.

2.4.1   ATM Protocol Overview

This section describes the basic concepts of ATM, ATM signaling, and ATM addressing; service classes that are offered by ATM; and ATM Adaptation Layer Protocols.

One of the advantages of ATM is its support for guaranteed QoS in connections to support delay and data loss requirements of user applications. A node requesting a connection can request a certain QoS from the network and be assured that the network will deliver that QoS for the life of the connection. Such connections are categorized into four ATM service categories: CBR, real-time and non-real-time variable bit rate (rt-VBR, nrt-VBR), available bit rate (ABR), and unspecified bit rate (UBR), depending on the nature of the QoS guarantee desired and the characteristics of the expected traffic. These ATM service categories are described in Section 2.4.6. Depending on the type of ATM service requested, the network is expected to deliver guarantees on the mix of QoS elements that specified at connection setup (e.g., cell loss ratio, cell delay, and cell delay variation). ATM QoS is covered in Chapter 8 in detail.

ATM is a connection-oriented protocol, which means that a connection must be established before a user can transmit data. A connection is identified by a virtual path identifier (VPI) and virtual channel identifier (VCI), as shown in Figure 2.10. A virtual path (VP) is a bundle of virtual channels (VCs) that are switched transparently across the ATM network on the basis of the common VPI. VPI and VCI have only local significance across a physical link and are remapped, as appropriate, at each switch.

Switching of ATM cells is performed in hardware based on the VPI/VCI fields of each cell. Switching performed on the VPI only is called a virtual path connection (VPC) (see Figure 2.10), while switching performed on both the VPI/VCI values is called a virtual channel connection (VCC) (see Figure 2.11). An ATM device may either be an endpoint or a connecting point for a VP or a VC. VPCs and VCCs exist between endpoints as shown in Figures 2.10 and 2.11, respectively. VP switching makes sense only when source and destination have a large volume of information to exchange. The VCIs within the VP remains unchanged and is transparent to the switching network, as all VCs will have the same VCI number at the ingress and egress side of the VP (end-to-end). VC switching makes sense when small, individual connections are necessary to maintain the required QoS (e.g., voice). The QoS is maintained throughout the switched network on a per-VC basis. Intermediate switches may modify both VPI and VCI values at each switch along the VCs route.

ATM PVCs and SVCs are described in Sections 2.4.2 and 2.4.3, respectively.

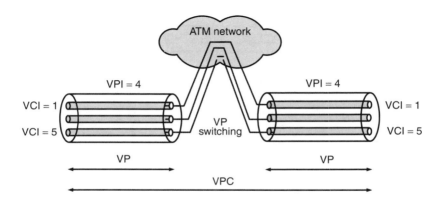

Figure 2.10 ATM VP switching.

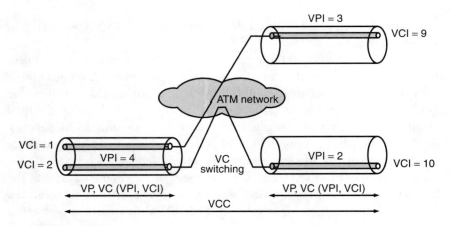

Figure 2.11 ATM VC switching.

As with Frame Relay, there are two interface standards defining the configurations of ATM networks. The UNI exists between the user equipment [i.e., end systems (ESs)] and switches. The NNI exists between two network switches.

Figure 2.12 shows the format of the 53-byte ATM cell at the UNI. The cell header contains an address significant only to the local interface, composed of two parts: an 8-bit VPI and a 16-bit VCI. This cell header also contains a 4-bit generic flow control (GFC), 3-bit payload type identifier (PTI), and a 1-bit cell loss priority (CLP) indicator and an 8-bit header error check (HEC) field.

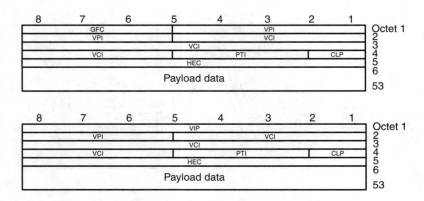

Figure 2.12 ATM UNI and NNI 53-byte cell formats.

Figure 2.12 shows the format of the 53-byte ATM cell at the NNI, identical to the UNI format with two exceptions. The NNI format uses the 4 bits used for the GFC at the UNI to increase the VPI field from 8 to 12 bits, allowing more VPs between network switches.

In theory, GFC supports simple multiplexing implementations at the UNI as defined in I.361 [21]. In practice, the current ATM Forum implementation agreements define only the uncontrolled mode, where the 4-bit GFC field is always coded as zeros. The GFC is part of the cell header, where, for example, a multiplexer that is connected to a set of terminals can control the terminals without using additional bandwidth. The GFC bits have different meanings depending on the direction on cell transmission. The protocol between the multiplexer and the terminals is asymmetric. Very few implementations support GFC, and it is supported only across an ATM UNI.

CLP indicates the loss priority of an individual cell. Either the end user or the network may set this bit. PTI discriminates between a cell payload carrying user information or one carrying management information. The HEC field provides error checking of the header by the physical layer.

2.4.2   ATM PVC Signaling

ATM PVCs are configured by the network operator: The set of switches between an ATM source and destination are programmed with the appropriate VPI/VCI values, traffic, and QoS parameters in both the forward and backward directions. Chapter 8 describes ATM traffic and QoS parameters in detail.

2.4.3   ATM SVC Signaling and Addressing

ATM SVCs are connections set up automatically through a signaling protocol. SVCs do not require the manual interaction needed to set up PVCs and, as such, are likely to be much more widely used. SVCs relieve some of the burden of provisioning VCs by making the process automatic. SVCs dynamically allocate VCs and are only active when data is being sent. SVCs are deallocated after a period of inactivity (timeout), saving valuable network resources. Unfortunately, the benefits of SVCs come with many drawbacks in the areas of billing, security, congestion, and troubleshooting. Billing systems for SVC-based services are more complex than for PVCs. SVCs are widely used in ISDN and X.25 networks but still add a level of complexity to the pricing equation. PVC-based networks are inherently more secure than SVC-based network. New VCs, when dynamically added, could potentially open up security holes that may not be acceptable in multicustomer networks. Since the call setup overhead requirements for a large-scale SVC network can overwhelm the resources of ATM switches at peak traffic periods, planning assumptions should be made regarding the amount of blocking allowed in the network. Problem isolation in an SVC-based network is much more difficult than it is with PVCs due to its transient nature, as there is no circuit to troubleshoot. Due to these limitations, most services are implemented using PVCs and will be for some time. In fact, PVCs are a familiar method of provisioning services, as they are widely used in Frame Relay networks, which are much larger and more complex than ATM networks.

SVC signaling is initiated by an ATM end system that desires to set up a connection through an ATM network; signaling packets are sent on a well-known virtual channel, addressed, for example, as VPI = 0, VCI = 5. The signaling is routed through the network, from switch to switch, setting up the new SVC connection identifiers as the signaling cell traverses the network toward its destination.

Signaling requests are routed between switches using address prefix tables within each switch, which precludes the need for a VC routing protocol. These tables are configured with the address prefixes that are reachable through each port on the switch. When a signaling request is received by a switch, the switch checks the destination ATM address against the prefix table and notes the port with the longest (i.e., best) prefix match. It then forwards the signaling request through that port to another switch. As the connection is set up along the path of the connection request, the SVC data also flows along this same path.

The ATM Forum UNI version 3.1 [22] signaling specification and ITU-T Recommendation Q.2931 [23] (called Q.93B) describe SVC signaling in detail. The user that originates the signaling (i.e., the calling party) initiates the call using a setup message containing information elements defining the called party number, as well as traffic and QoS parameters in both the forward and backward directions. The network responds to the caller's setup message with a call-proceeding message, as shown in Figure 2.13. The network routes the call (SVC) across intermediate switches by forwarding the setup message, reserving the required capacity and other resources at the switch, based on the information in the message. Eventually, the network routes the setup message to the destination called party number. The called user responds with the connect message, indicating the locally assigned VPI/VCI label values in the connection identifier (CI). The network propagates the connect message back to the caller. The users and the network

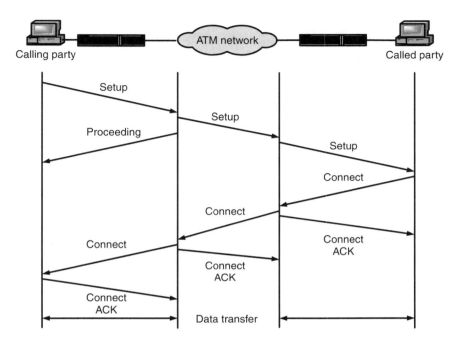

Figure 2.13 ATM SVC call setup procedures.

confirm the connect message using the connect-acknowledge message. As the end-to-end connection is now established, data cells can be sent between the calling and called users, following the path reserved by the setup message.

In order to send signaling messages such as setup, addressing capability is required. The telephone number-like E.164 addressing structure is the basis for addressing public ATM (B-ISDN) networks. Since E.164 addresses are public (and expensive) resources and cannot typically be used within private networks, the ATM Forum extended ATM addressing to include private networks. In developing a private network addressing scheme for UNI 3.0/3.1 [22], the ATM Forum defined an address format based on the syntax of an OSI network service access point (NSAP) address. However, ATM addresses are not NSAP addresses, despite their similar structure. While these addresses are commonly referred to as *NSAP addresses*, they are better described as ATM private network addresses or ATM end-point identifiers. These 20-byte NSAP format addresses are designed for use within private ATM networks, while public ATM networks typically use E.164 addresses that are formatted as defined by ITU-T.

2.4.4 ATM Layered Model

The ATM layered model consists of physical layer, ATM layer and ATM adaptation layer (AAL), as shown in Figure 2.14. This layered architecture is based on the B-ISDN protocol stack defined in ITU-T Recommendations I.321 [24] and I.363 [25]. The physical layer is divided into two sublayers: transmission convergence (TC) and physical medium dependent (PMD). The PMD sublayer interfaces with the actual electrical or optical transmission medium, detecting the optical signals, transferring bit timing, and passing the bit stream to and from the TC sublayer. The TC sublayer extracts and inserts ATM cells within either a plesiochronous or synchronous digital hierarchy (PDH or SDH) TDM frame passed to and from the ATM layer, respectively. The ATM layer performs multiplexing, switching, and control actions based on information in the ATM cell header and passes cells to, and accepts cells from, the AAL. The adaptation layer (as its name implies) adapts higher layer protocols, be they signaling or user information, to fixed-size ATM cells. The adaptation layer is further described in Section 2.4.5.

The ATM layer is fully independent of the physical medium used to transport the ATM cells. It provides the following functions:

- *Cell construction*—ITU-T Recommendation I.361 [21] defines the coding of ATM cells. The ATM layer is responsible for constructing the 5-byte ATM cell header for transmission via the physical layer.

- *Cell reception and header validation*—The ATM layer is responsible for receiving ATM cells and forwarding the 48-byte payload to the ATM adaptation layer. Header validation is performed on the VPI, VCI, and PTI fields.

- *Cell relaying, forwarding, and copying using the VPI/VCI*—The heart of ATM is the use of the VPI and VCI for relaying or switching. For VC switching, ATM cells arriving are switched based on the VPI and VCI in the ATM cell header. For VP switching, ATM cells are

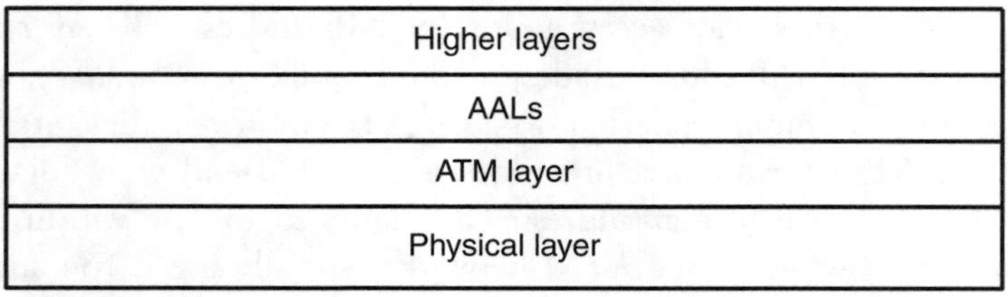

Higher layers
AALs
ATM layer
Physical layer

Figure 2.14 ATM layered model.

switched based on the VPI only. For VP switching, the VCI in the ATM cell header remains unchanged for all VCs in the VP. Multicasting is done in PMP connections, where each ATM cell is copied (multiple copies) and sent out on different VCs. Video broadcasting is an example where video is broadcast from a single source and relayed to several destinations.

- *Cell multiplexing and demultiplexing using the VPI/VCI*—ATM also performs multiplexing and demultiplexing of multiple logical connections with different QoS requirements.

- *Cell payload type discrimination*—ITU-T Recommendation I.361 defines the PTI encoding that identifies various cell types in ATM. For example, cells that carry user information and those that carry ATM management information [operation and management (OAM)] have different payload types. Payload types carrying user information may indicate congestion by the explicit forward congestion indication (EFCI) bit. This field also indicates the last cell of an AAL5 payload when packets of information are transmitted using AAL5 (see Section 2.4.5).

- *Interpretation of predefined reserved header values*—The ATM layer performs the identification and processing of preassigned, reserved header vales. ITU-T Recommendation I.361 defines these preassigned values for the ATM UNI interface. ITU-T Recommendation I.432 [26] defines specific values and meanings for the physical layer OAM cells.

- *CLP bit processing*—A value of 0 in the CLP field specifies highest priority (i.e., the network is least likely to discard CLP = 0 cells in the event of congestion). A value of 1 in the CLP means that the cell has low priority (i.e., the network may selectively discard CLP = 1 cells during congested intervals in order to maintain a low loss rate for the high-priority cells).

- *Multiple QoS classes*—Many data applications respond to delay and loss through retransmission strategies designed to provide guaranteed delivery. Applications such as voice and constant bit rate video are much less tolerant to delay and loss than variable bit rate applications. QoS classes that guarantee delay and cell loss are covered in detail in Chapter 8.

2.4.5   ATM Adaptation Layer

The AAL isolates higher layers from the specific characteristics of the ATM layer. AAL converges or adapts the user traffic to the ATM cell-based network to support different classes of traffic, such as voice, data, and video. The AAL converts and aggregates traffic into standard formats as defined by the AAL protocols described later in this section.

Different types of AALs support different application traffic requirements. The various AALs have vastly different protocol data unit (PDU) lengths. AAL1 and AAL2 have shorter PDU lengths to support real-time service. AAL3/4 and AAL5 support traditional packet data services by carrying payloads ranging from a single byte up to 65,535 bytes. AAL5 also supports multiple logical channels over a single ATM VPI/VCI. The following describes the different layer types in more detail:

- *ATM adaptation layer type 1 (AAL1):* Defined in ITU-T Recommendation I.363.1 [25], AAL1 supports CBR traffic, such as real-time video and DS1/DS3 CES that interwork with TDM protocols. AAL1 transfers service data units received from a source at a constant source bit rate and delivers them at the same bit rate to the destination. Chapter 7 describes in detail the application of the CES interworking using AAL1.

- *ATM adaptation layer type 2 (AAL2):* Defined in ITU-T Recommendation I.363.2 [25], AAL2 supports VBR traffic such as packetized audio-video. The function of the AAL2 protocol is to minimize delay and improve efficiency for real-time voice and video.

- *ATM adaptation layer type 3/4 (AAL3/4):* Defined in ITU-T Recommendation I.363.3 [25], AAL3/4 supports VBR traffic for such applications as SMDS. The initial developments for AAL3 (connection) and AAL4 (connectionless) were so similar that ITU-T merged them into a single AAL3/4.

- *ATM adaptation layer type 5 (AAL5):* Defined in ITU-T Recommendation I.363.5 [25], AAL5 is a leaner version of AAL3/4 that supports Frame Relay and IP. AAL5 is known as the simple efficient adaptation layer (SEAL) because the protocol is far less complicated than any of the other AALs. Support for both connection-oriented and connectionless VBR traffic is provided by AAL5. Most data-oriented applications use AAL5. Chapter 6 describes in detail Frame Relay interworking using AAL5.

- *ATM adaptation layer type 0 (AAL0):* Not defined in any specification, AAL0 does not segment the application data. Data must fit exactly into one 48-byte cell. AAL0 is used to convey operations, administration, and maintenance (OAM) information defined specifically to manage the ATM network as described in Section 2.4.9.

2.4.6 ATM Service Categories

The ATM Forum Traffic Management 4.1 [27] specification defines the following ATM layer service categories, labeled by bit rate and implied QoS:

The *ATM CBR* service category supports applications that require a static amount of bandwidth that remains continuously available throughout the connection lifetime. The amount of bandwidth is characterized by a maximum bit rate. For applications such as voice, constant-bit-rate video, and CES, the delay caused by transmission and queuing can significantly reduce the value of the data that is being received by these applications. Similarly, any loss of data due to transmission errors and congestion (buffer overflows) can also contribute to service degradation. This service is best described as delay and loss sensitive service.

The *ATM rt-VBR* service category supports applications that are intended for bursty real-time applications and those requiring tightly constrained delay and delay variation. Applications include packetized voice and variable-bit-rate video. These applications do not generate a constant bit stream but are still delay and loss-sensitive.

The *ATM nrt-VBR* service category supports applications that have no constraints on delay and delay variation but still have variable bit rate and bursty traffic characteristics. This class of applications can tolerate a low cell loss because retransmission by the applications can recover the lost data without adversely affecting service delivery to the user. Applications include packet data transfers, terminal sessions, and file transfers.

The *ATM UBR* service category, known as the *best-effort* service, requires neither tightly constrained delay nor delay variation. UBR applications require no specific QoS or guaranteed throughput whatsoever. The Internet, LAN emulation, and IP over ATM are examples of UBR applications.

The *ATM ABR* service category supports applications that can change their transmission rate in response to rate-based network feedback in the context of closed-loop flow control. The primary aim of ABR is to dynamically provide access to capacity currently not in use by other service categories to users who can adjust their transmission rate in response to feedback. In exchange, the network provides a service with very low loss. Real-time applications are not suitable, as this service does not guarantee any bounded delay variation. Applications that are suitable for ABR include LAN interconnection, high-performance file transfer, database archival, non-time-sensitive traffic, and Web browsing.

2.4.7   AAL Service Category Mapping

Figure 2.15 identifies the applications that use particular combinations of AALs and ATM service categories. In summary, AAL1 interworks with TDM protocols using circuit emulation (i.e., CBR). AAL2 provides the infrastructure to support VBR voice and video. Most data-oriented applications use AAL5, which includes an early specification for VOD defined by the ATM Forum in advance of AAL2 standardization. Early LAN and internetworking protocols (i.e., LANE and classical IP over ATM) use only the UBR, also known as the best effort, service category. The ATM Forum's LANE specification version 2.0 [5] provides access to all ATM service categories. There are also few applications that use the ATM layer directly without AALs.

2.4.8   ILMI Protocol

Integrated local management interface (ILMI) functions provide configuration, status, and control information about physical and ATM layer parameters of the ATM interface. The ATM Forum's ILMI specification (af-ilmi-0065.000) [28] and ATM UNI version 3.1 [22] define the ILMI interface in detail. The ILMI protocol operates on both ATM UNI and NNI interfaces using dedicated VPI and VCI values (VPI = 0, VCI = 16).

To facilitate administration and configuration of ATM addresses into ATM end systems across UNI, the ATM Forum defined an address registration mechanism using the ILMI protocol. This mechanism not only facilitates automated addresses, but may also be extended, in the future, to allow for auto configuration of other information (such as higher-layer addresses and server addresses).

Although UNI and NNI cell formats specify 8 or 12 VPI bits and 16 VCI bits, real ATM systems typically use fewer bits. VPI/VCI bits supported by interconnected devices must be identical for interoperability. One way to facilitate VPI/VCI interoperability is to use ATM Forum's ILMI, which allows each system to query others about the number of bits supported.

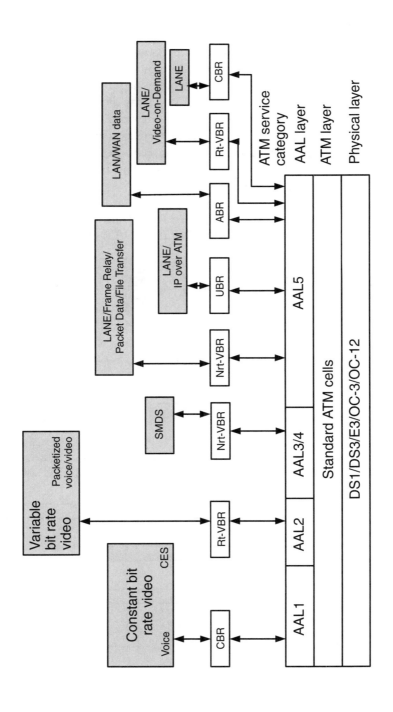

Figure 2.15 Mapping of ATM service categories.

2.4.9 Operations and Maintenance Cells

ATM OAM cells are used to exchange operations information between ATM nodes, including customer premises equipment. End-to-end management functions are carried transparently through the ATM switch and made available to the user using end-to-end OAM cells as shown in Figure 2.16. End-to-End OAM cells pass unmodified by all intermediate nodes. The content of these cells may be monitored by any node in the path and removed by the endpoint. Segment management functions are carried by segment OAM cells and removed at the end of a segment. A segment is defined as a single VP or VC link as shown in Figure 2.16.

The I.610 [29] ATM layer management standard defines OAM cells with a special format. OAM cells are defined for VP and VC connection for either an end-to-end or a segment. At the VP level, OAM cells have the same

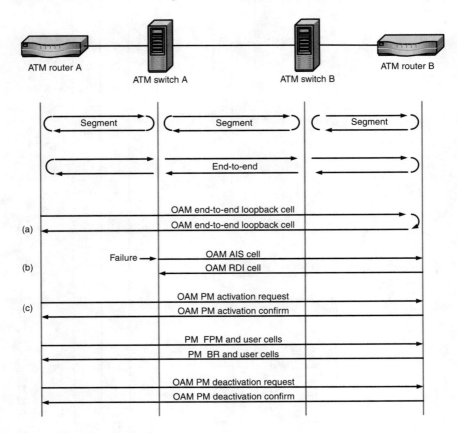

Figure 2.16(a–c) ATM OAM end-to-end and segment flows.

VPI values as the user-data cells but are identified by preassigned VCI values whether the flow is either end-to-end VCI = 3 or segment VCI = 4 (see Figure 2.17). At the VC level, OAM cells have the same VPI/VCI values as the user-data cells but are identified by preassigned code points of the PTI, which differentiates between the end-to-end PTI = 100 and segment PTI = 101 (see Figure 2.17).

ATM OAM cells are used within an ATM network for the following management functions:

1. *Fault management*—These determine when there is a failure, notifying other elements of the connection regarding the failure and providing the means to diagnose and isolate the failure.

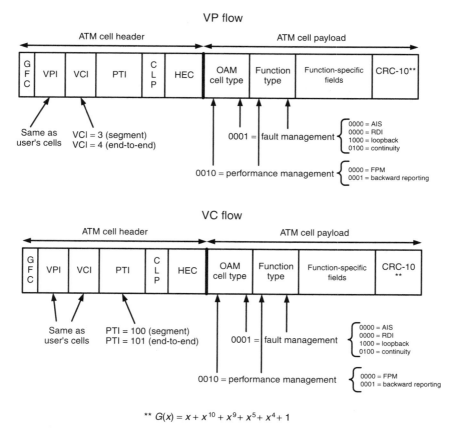

$$** \ G(x) = x + x^{10} + x^9 + x^5 + x^4 + 1$$

Figure 2.17 Common OAM cell format. (*From:* ATM UNI version 3.1, af-uni-0010.002, 1994.)

- *Loopback*—These are used to verify connectivity and help diagnose problems due to misconfiguration of VPI or VCI translations in a configured VP or VC. Figure 2.16(a) illustrates an example of an end-to-end loopback initiated at ATM router A, which inserts an OAM end-to-end loopback cell. The ATM router B extracts the loopback cell and transmits the loopback cell in the opposite direction as shown in the figure. Note that the intermediate ATM switches are extracting and processing every OAM cell, which results in ATM switch A and B transparently conveying the end-to-end loopback cell between ATM router A and B. Eventually, ATM router A extracts the loopback cell. The correlation tag within the payload of the end-to-end OAM cell is used to match the sent and received OAM cell, verifying continuity of the VP (or VC) from ATM router A to B.

- *Alarm indication and remote defect indication*—Alarm surveillance involves detection, generation, and propagation of VP and VC failures (indications). In analogy with the SONET physical layer, the failure indications are of two types: alarm indication signal (AIS) and remote defect indication (RDI). AIS/RDI is used to detect and identify to the connection endpoint that a failure has occurred. Figure 2.16(b) illustrates where a failure at ATM switch A generates an OAM end-to-end AIS cell toward ATM switch B and beyond. This indicates in the downstream direction that a failure in the upstream has occurred. The ATM router B generates an OAM end-to-end RDI cell in the opposite direction as indicated by the figure. The AIS cells are normally generated every second until the recovery of the failure. The failure can be detected at the VP or VC level.

- *Continuity check*—An endpoint sends a cell periodically at some predetermined interval if no other traffic is sent on the VC so that the connecting points and other endpoints can distinguish between a connection that is idle and one that has failed. The ATM Forum currently does not specify support for the continuity check for interoperability at the ATM UNI.

2. *Performance management*—Measurement of QoS is done by inserting OAM cells on a configured VP or VC and monitoring blocks of user cells. The user traffic is not impacted by the measurements as the OAM cells are inserted nonobtrusively. The monitoring will detect errored blocks and loss/misinsertion of cells within a defined

block of user cells. Figure 2.16(c) illustrates the activation and deactivation process that is involved in measuring performance between ATM router A and B. ATM router A transmits an activation request toward ATM router B. If ATM router B can comply with the request, it will return activation confirm; if not, activation denied will be returned. Activation and deactivation allows performance measurement to be performed on a selected VP or VC. The performance monitoring (PM) block size (128, 256, 512, 1,024, ... 32,768) information is sent in the activation request in the direction A to B. The activation confirm will also contain a PM block size in the direction B to A, as this may be different. Forward performance monitoring (FPM) cells are inserted for the VP or VC after every N user cells (N is the PM block size) at ATM router A. ATM router A will send in the FPM cell, the total number of user cells since the last FPM cell and block error detection code computed over all of the user cells since the last FPM cell (used for error-rate estimation). ATM router B will extract the FPM cell, calculate the receive counts, and recompute the 16-bit parity for comparison with the value received in the FPM cell computed by ATM router A. The PM results are sent by ATM router B, using the backward reporting (BR) cell that contains the block error result, which is the number of errored parity bits detected across user cells and the number of received user cells. Note that the BR cell contains the results for the cells in the preceding user block. The calculation of QoS cell loss ratio (CLR), cell error ratio (CER), and cell transfer delay (CTD) can be done by ATM router A, based on these results. Finally ATM router A will deactivate the PM session as shown in Figure 2.16(c).

2.5 Summary

We have covered the protocol basics of IP, ATM, and Frame Relay in this chapter. IP and ATM and, to a lesser extent, Frame Relay, will form the key protocols that will govern the different interworking arrangements that will exist in future networks. In addition to implementing functional requirements, internetworking arrangements will present issues and challenges when other value-added provisions (e.g., security protocols) or physical media transmission characteristics (i.e., wireless access segment) are taken into account.

References

[1] Heinanen, J., "Multiprotocol Encapsulation over ATM Adaptation Layer 5," RFC 1483, July 1993.

[2] Laubach, M., and J. Halpern, "Classical IP and ARP over ATM," RFC 1577/2225, April 1998.

[3] Luciani, J., et al., "Next-Hop Resolution Protocol (NHRP)," RFC 2332, April 1998.

[4] The ATM Forum Technical Committee, "LAN Emulation over ATM Version 1.0," af-lane-0021.000, January 1995.

[5] The ATM Forum Technical Committee, "LAN Emulation over ATM Version 2.0," af-lane-0084.000, July 1997.

[6] The ATM Forum Technical Committee, "Multi-Protocol over ATM Version 1.1," af-mpoa-0114.000, May 1999.

[7] Nichols, K., et al., "Definition of the Differentiated Services Field (DS Field) in the IPv4 and IPv6 Headers (DiffServ)," RFC 2474, 1998.

[8] Braden, R., et al., "Resource ReSerVation Protocol (RSVP)," RFC 2205, September 1997.

[9] Davie, B., et al., "MPLS Using LDP and ATM VC Switching," RFC 3035, January 2001.

[10] Kent, S., and R. Atkinson, "Security Architecture for the Internet Protocol (IPsec)," RFC 2401, November 1998.

[11] IEEE Standards Board, "Overview and Architecture," IEEE 802, May 31, 1990. (Note: This standard provides an overview to the family of IEEE 802 standards. This document forms part of the 802.1 scope of work).

[12] Rekhter, Y., B. Moskowitz, and D. Karrenberg, "Address Allocation for Private Internets," RFC 1918, February 1996.

[13] Lougheed, K., Y. Rekhter, and T. J. Watson, "A Border Gateway Protocol 3 (BGP-3)," RFC 1267, October 1991.

[14] Droms, R., "Dynamic Host Configuration Protocol (DHCP)," RFC 1531, October 1993.

[15] Gilmore, J., and B. Croft, "Bootstrap Protocol (BootP)," RFC 951, September 1985.

[16] ITU-T Recommendation Q.922, "ISDN Data Link Layer Specification for Frame Mode Bearer Services," February 1992.

[17] ITU-T Recommendation Q.933 bis, "Digital Subscriber Signaling System No. 1," October 1995.

[18] ANSI T1.617a-1994, Integrated Services Digital Network (ISDN)—Signaling Specification for Frame Relay Bearer Service for Digital Subscriber Signaling System Number 1 (DSS1), 1994.

[19] Frame Relay Forum Implementation Agreement FRF.1.1, "User-to-Network (UNI) Implementation Agreement," January 1996.

[20] ITU Recommendation I.370, "Congestion Management for the ISDN Frame Relaying Bearer Service," 1991.

[21] ITU-T Recommendation I.361, "B-ISDN ATM Layer Specification," February 1999.

[22] The ATM Forum Technical Committee, "User-to-Network Interface (UNI) Specification Version 3.1," af-uni-0010.002, 1994.

[23] ITU-T Recommendation Q.2931, "B-ISDN Application Protocols for Access Signaling," February 1995, Amendment 1, June 1997.

[24] ITU-T Recommendation I.321, "B-ISDN Protocol Reference Model and Its Application," April 1991.

[25] ITU-T Recommendation I.363, "B-ISDN ATM Adaptation Layer (AAL) Specification," March 1993.

[26] ITU-T Recommendation I.432, "B-ISDN User-Network Interface—Physical Layer Specification," February 1999.

[27] The ATM Forum Technical Committee, "Traffic Management Specification Version 4.1."

[28] The ATM Forum Technical Committee, "ILMI Specification Version 4.0," af-ilmi-0065.0000.

[29] ITU-T Recommendation I.610, "B-ISDN Operation and Maintenance Principles and Functions," November 1995.

Selected Bibliography

ITU-T Recommendation I.356, "B-ISDN ATM Cell Transfer Performance," March 2000.

ITU-T Recommendation I.311, "B-ISDN General Network Concepts," August 1996.

ITU-T Recommendation I.413/ANSI T1.624, "B-ISDN User Network Interface," March 1993.

ANSI T1.627 Telecommunications Broadband ISDN, "ATM Layer Functionality and Specification," 1993.

ITU-T Recommendation I.113, "Vocabulary of Terms for Broadband Aspects of ISDN," June 1997.

Huitema, C., "IPv6: The New Internet Protocol," Upper Saddle River, NJ: Prentice Hall, 1996.7

3

Wireless Segment in Internetworking Architecture

Broadband access services, especially wireless access services, have become important components in CLECs' efforts to use complementary technologies in a layered architecture to provide flexible service offerings. This chapter expands on key wireless technologies and explains their role as integral components in internetworking arrangements within a multitiered national network. This chapter also provides background on the backhaul networks to which wireless segments are connected.

3.1 Internetworking Scenario Overview

Figure 3.1 [1] illustrates a CLEC's continental network with typical layered architecture. The backbone network ring involves SONET services based on leased or owned fiber. The second tier is based on regional fiber networks. The third tier consists of metropolitan-area networks (MAN), which are supported by backbone and regional networks. This layered architecture provides the framework for supporting broadband access at the network edge, using access technologies like DSL, ATM, optical transmission and wireless schemes such as LMDS, multichannel multipoint distribution system (MMDS), point-to-point wireless, and wireless local loop.

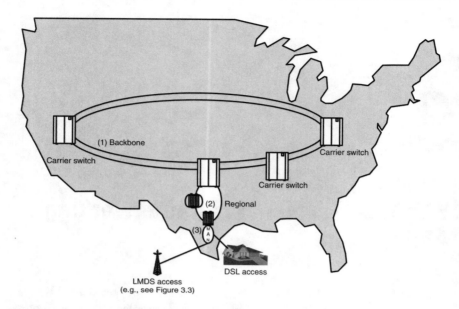

Figure 3.1 Three-tier CLEC network architecture. (*From:* P-COM [1], 1994.)

The physical media supporting this network is based on high-performance fiber and popular multiplexing technologies [for example, wave division multiplexing (WDM)]. The layer 2 communication stack can consist of ATM, IP, SONET, or a combination of these technologies. This section considers the described architecture and the level 1 and 2 technologies that serve as enablers for alternative wireless access technologies.

Figure 3.2 identifies different types of customers in a typical metropolitan environment, their typical bandwidth requirements, and the access technologies that can meet these requirements.

In large metro areas, bandwidth needs for large to medium businesses in multitenant buildings range from fractional-T1 and ATM25 to T3 or OC-3 rates. While ILECs may prefer to use nonwireless technologies to deliver bandwidth, last-mile issues render wireless technologies efficient and cost-effective, allowing CLECs to expand their customer base by offering alternative access methods.

Figure 3.3 illustrates how various wireless technologies can be used in conjunction with fiber networks supporting ATM, IP, or packet over SONET protocols.

In a typical optical ring providing the network for a city, segment (A) represents a broadband wireless point-to-point link installed, for example, to

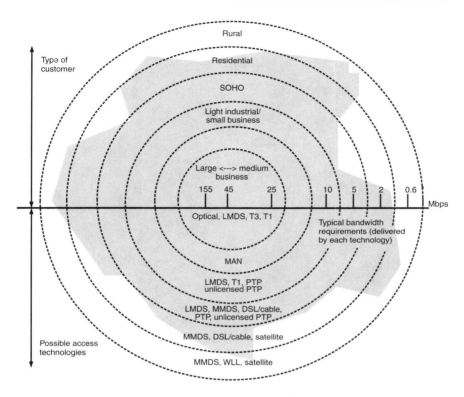

Figure 3.2 Access technologies and bandwidth requirements [1]. (*From:* P-COM, 1994.)

overcome a physical barrier that blocks signal or adds cost. Radio spur (B), topologically similar to an optical spur, may be used when buildings are not reachable by optical cable or to collect traffic data from mobile base stations not reachable from an LMDS station. The cable subring extending from (C) is a conventional means for extending access to a fiber ring. LMDS fixed cell (D) uses a base station in a central tall building to serve several remote stations at customer buildings. LMDS base stations generally use ATM technology to connect to a switch, enabling these networks to provide voice, data, and video services with guaranteed QoS. Some LMDS products provide IP-based services for more data-centric applications, confronting more challenging QoS issues. Different configurations of LMDS systems are covered in depth in later sections of this chapter. MMDS cells like (E) can be linked to the backbone network using Frame Relay access protocol. LMDS and MMDS are similar wireless technologies that use different frequency bands and therefore deliver different bandwidths to different service areas. A radio

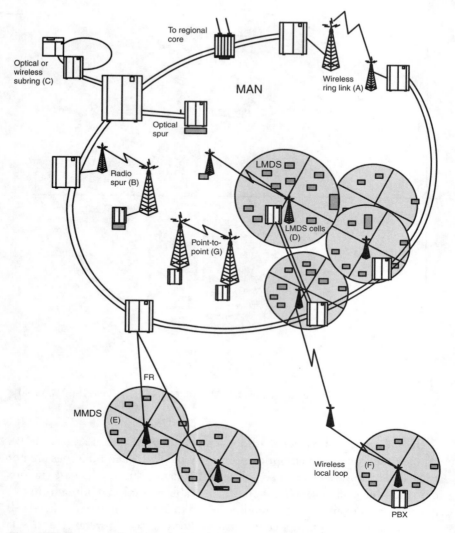

Figure 3.3 Access technologies in metropolitan networks [1]. (*From:* P-COM, 1994.)

relay link (F) that connects to an LMDS node through a repeater link provides connectivity to a wireless local loop (WLL) network. The WLL provides basic wireless telephony service over a wide area using VHF and UHF links. A private, point-to-point network (G) is implemented with broadband wireless radio links. This network simply joins two buildings that have no need for public networking or other services.

These examples illustrate the importance of wireless segments in evolving MAN architectures and highlight the importance of interworking of protocols supporting the wireless components and interfacing networks. ATM, IP, and Frame Relay protocols figure prominently in these interworking scenarios. Implementation of interworking arrangements according to existing standards that facilitate seamless interconnections and guarantee QoS to users present formidable challenges to network designers.

3.2 UNI

Service providers are using ATM or packet over SONET in their network cores, while customers are generally considering IP as the protocol of choice within enterprise networks. In addition, interworking of Frame Relay with ATM and IP also plays a role.

User interfaces to the wireless access network are therefore likely to be Frame Relay (may be carrying IP), simple IP, or ATM (carrying other protocols). The wireless far-end is likely to be ATM, with the customer far-end being Frame Relay [2], IP, or ATM. Wireless access devices also provide various forms of DSL access multiplexing, allowing ATM output to be carried over wireless segments, for reconversion into DSL on the last mile.

The wireless segment must therefore provide interworking between these core and edge protocols. As we will see in Chapter 4, several existing access devices can perform the interworking functionality. Increasingly, remote stations supporting the wireless segment are being designed to incorporate access device functionality.

3.3 Overview of LMDS, MMDS, Point-to-Point, and WLL

LMDS, or local multipoint communication service (LMCS), as the technology is known in Canada, is a wireless, two-way broadband technology designed to allow network integrators and communication service providers to quickly and inexpensively bring a wide range of high-value, quality services to homes and businesses. LMDS can be contrasted with WLL, narrowband technology primarily used to provide basic telephony service. Broadband LMDS can deliver services that include interactive television, high-speed data for the Internet, and multiple voice channels, to name a few. LMDS is designed to deliver data through the air at rates up to 155 Mbps (compared to typical cell phone voice calls using a mere 64 Kbps, or 8 Kbps in compressed digital systems).

Of the many technologies developed for high-speed wireless access, LMDS offers an ideal way to break through the local-access bottleneck. Multipoint radio technology, combined with the appropriate protocol and access method, gives LMDS tremendous potential. LMDS is ideal for transporting VoIP and switched video, as well as computer data. By official definition, LMDS only applies to services operating in two frequency bands: 28 GHz and 31 GHz. But the term has also been applied to analogous services operating at 24 GHz, 38 GHz, and 39 GHz in the United States and 26 GHz internationally.

Other wireless technologies use low-frequency signals in the lower RF spectrum, with enough power to penetrate buildings over long distances. LMDS uses low-powered, high-frequency signals over a short distance. LMDS systems are cellular because they send these very high-frequency signals over short line-of-sight distances. These cells are typically spaced 4 to 5 km (2.5–3.1 mi) apart. LMDS cell layout determines the cost of building transmitters and the number of households covered.

Unlike a mobile phone, which a user can move from cell to cell, the transceiver of an LMDS customer has a fixed location and remains within a single cell. LMDS systems consist of multicell configuration distribution systems with return path capability within the assigned spectrum. Generally, each cell will contain a centrally located transmitter (base station), multiple receivers or transceivers, and point-to-point links interconnecting the cell with a central processing center or other cells.

Due to small cell size, the same spectrum is often repeated in many cells. Since each cell repeats some portion of the spectrum in some form, this multiplies the aggregate capacity of LMDS. This fundamental spectrum reuse scheme is forced by the propagation characteristics of high-frequency systems such as LMDS. Another method of capacity multiplication is to sectorize the transmission pattern. In one simple version, instead of transmitting channel A on a 360-degree pattern, and channel B on a 360-degree pattern (effective coverage of two channels), channel A and B are alternated in pie-shaped sectors. If, for example, 12 30-degree sectors are used, 6 channel As and 6 channel Bs are created, each capable of carrying completely separate data. The effective coverage of this example is 12 channels, or more if sectors are allowed to overlap. All omnidirectional wireless systems apply this frequency reuse method for outbound transmission, inbound reception, or both.

LMDS cell size is limited by *rain fade,* distortions of the signal caused by raindrops scattering and absorbing the millimeter waves (the same process that heats food in a microwave oven). Also, walls, hills, and even leafy trees block, reflect, and distort the signal, creating significant shadow areas for a single transmitter. ATM is used extensively in wide-area LMDS networks, allowing a mixture of data types to be interleaved. Thus, a high-quality voice service can run concurrently over the same LMDS data stream as Internet, data, and video applications.

MMDS, occupying frequencies located in the 2.1–2.7-GHz band, is another option to deliver broadband wireless services. MMDS is attractive for the small office/home office (SOHO) market that requires low bandwidth (up to 3 Mbps) over a much larger area (up to 50 km). MMDS frequencies have traditionally been used to provide one-way, analog wireless cable TV broadcast service. As such, the MMDS industry is more widely known as the *wireless cable industry*. Unlike LMDS, MMDS provides broadband digital data and TV directly into the home.

WLL is a fixed wireless technology for the local loop that uses a wireless link to connect subscribers to a local exchange in place of conventional copper cable. With WLL, there is radio equipment at the customer premises, which is connected to the inside wiring of a residence or commercial dwelling. The radio equipment communicates with the base tranceiver station (BTS) cell site just like any mobile device (such as a mobile phone). Distinguishing WLL from wireless for mobile networks is the fact that WLL is a fixed wireless solution, not allowing for mobility. Since a WLL system serves as the access line for fixed telephone sets, it must provide the same level of quality as conventional telephone systems with respect to such aspects as speech quality, connection delay, and speech delay. Like satellite, WLL (also known as fixed-point wireless) is offered primarily to businesses and homes in areas where the infrastructure is not in place. WLL systems operate at frequencies of 800 MHz and 1.9 GHz.

A wireless point-to-point link consists of two microwave radio systems linking two locations (e.g., two buildings). Instead of linking the two locations with leased lines, dial-up connections, or fiber optics, it is cost-effective to use a wireless radio link. For example, a 2.4-GHz radio point-to-point link between two buildings requires line of site, which enables the link to operate error free.

3.4 Broadband Wireless LMDS Architecture

Various network architectures are possible within a LMDS system design. Point-to-point LMDS networks offer high-speed dedicated links between each base station and remote station. PMP as shown in Figure 3.4, LMDS networks offer high-capacity local access that is less capital-intensive than a wireline solution. In this configuration, one base station is linked with several remote stations. The signals are transmitted in a PMP (or broadcast) method; the wireless return path from the remote station to the base station is point-to-point.

The base station is where the conversion from a fiber to a wireless infrastructure occurs. One approach for the design of the base station is to simply provide connections to the fiber infrastructure. This forces all traffic to terminate in ATM switches or central office (CO) equipment somewhere in the fiber infrastructure. In this architecture, two customers connected to the same base station who wish to communicate with each other will be connected at the ATM switch or CO equipment, as shown in Figure 3.5. On the other hand, if local switching is present at the base station, the two customers can be connected at the base station without entering the fiber infrastructure.

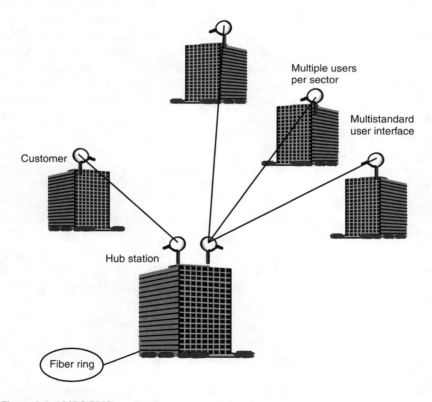

Figure 3.4 LMDS PMP applications.

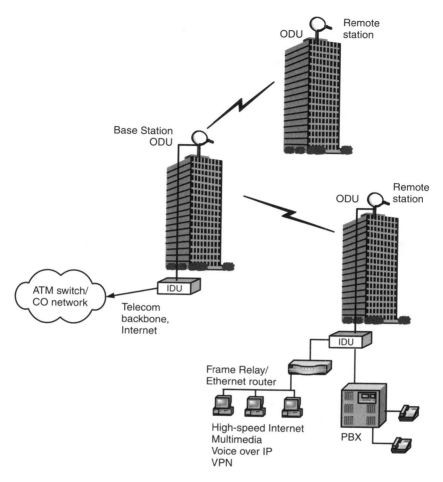

Figure 3.5 LMDS base station and remote stations.

3.4.1 LMDS Base Station

Physically, the base station at the customer site consists of two entities: the outdoor unit (ODU) and the indoor unit (IDU).

The base station's ODU enables wireless signal transmission and reception. A common design puts the remote station's ODUs on rooftops, to get a good line of sight to the base station's ODU transceiver.

The base station's IDU provides the basic network gateway for connecting wireline network traffic to the LMDS bandwidth. The LMDS base station is equivalent to base station digital equipment in a cellular network.

The base station provides processing, multiplexing/demultiplexing, compression, error detection, encoding/decoding, routing, and modulation/demodulation. Once a base station is installed, new customers can be added in a matter of hours.

Centralized base stations transmit via omnidirectional or broadbeam sector antennas with beamwidths of 45, 90, or 180 degrees. The base station's transmitter links to subscriber locations equipped with roof-mounted 10–12-in directional ODU antennas.

3.4.2 LMDS Remote Station

A remote station located at the customer premises serves as the gateway between the RF component and in-building CPE, as shown in Figure 3.6. Physically, the remote station also consists of ODU and IDU entities.

The IDU provides industry-standard UNIs such as IP, pulse code modulated (PCM) voice, Frame Relay, and native ATM. System interfaces

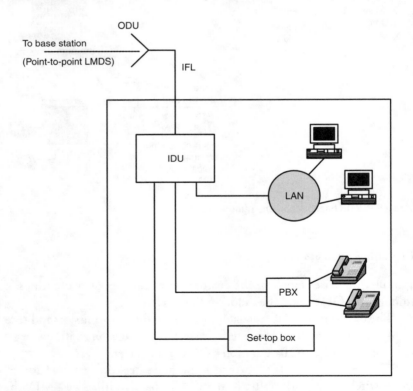

Figure 3.6 LMDS remote station.

are compatible with existing PBXs, switches, routers, LANs, and bridges, providing a highly adaptable solution to meet varying network design requirements.

The IDU at the remote station is designed to be scalable and flexible and chassis-based. The IDU can be configured with 10/100 BaseT, analog voice, structured and unstructured T1/E1, T3/E3, and OC-3/STM-1 fiber communications, to name a few interworking applications. The remote station IDU communicates with the base station through an ODU (rooftop), forming a part of the PMP LMDS network. IDUs and ODUs are connected via a single intermediate frequency link (IFL) coaxial cable.

3.4.3 RF Subsystem: U.S. and European Frequency Bands

In the United States, regional Bell operating companies (RBOCs) have built a wired high-speed infrastructure for data transmission. In order to create viable opportunities for wireless competition, the FCC enhanced the capacity of existing spectrum licenses to promote bidirectional transport without requiring receive site licenses.

The high capacity of LMDS is possible because it operates in a large, previously unallocated expanse of the electromagnetic spectrum. In the United States, the Federal Communications Commission (FCC) allocated a total bandwidth of about 1.3 GHz in the millimeter waveband at frequencies of about 28 GHz. In other countries, depending on the local licensing regulations, LMDS operates at 10.5 or 26 GHz, as shown in Figure 3.7.

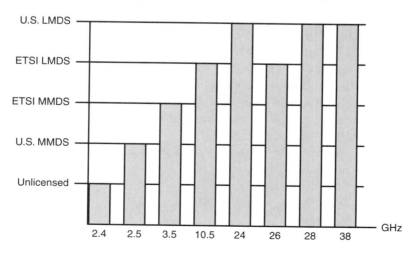

Figure 3.7 LMDS and MMDS frequencies.

In 1998, the FCC carried out an auction of spectrum in the 28–31-GHz range. In each geographical area, the FCC auctioned an A block (with a bandwidth of 1,150 MHz) and a B block (with bandwidth of 150 MHz). LMDS licensees are able to offer local-exchange telephone service, Internet access, and other broadband services in their purchased blocks.

The 10 GHz band, with frequency range 10.15 to 10.65 GHz, is primarily used in international markets. Frequencies are allocated in the United States by the FCC and in other countries by local regulatory bodies as shown in Figure 3.8.

Block A consists of frequencies in the 27,500–28,350-MHz, 29,100–29,250-MHz, and 31,075–31,225-MHz bands.

Block B is comprised of spectrum in the 31,000–31,075-MHz and 31,225–31,300-MHz bands. A Block license holders have a 1,150-MHz slice in the 28- and 31-GHz bands, which is large enough to support cable television (CATV), local telephony, and broadband access simultaneously to users in each cell. The 38-GHz band in the 38.6–40.0-GHz range is used in the United States, where it is primarily licensed to Winstar and Advanced Radio Telecommunications (ARTT). Europeans use the 40-GHz band for LMDS.

3.4.4   RF Subsystem: Modulation Techniques

Signals from the voice, video, and data multiplexing system are modulated before wireless transmission occurs. Similarly, traffic from the microwave receiver is demodulated before wireline transmission. Modulation is the conversion of bits to hertz. Modulation methods for LMDS systems are generally separated into phase shift (PSK) and amplitude modulation (AM). The modulation options for TDMA and FDMA access methods are almost the same. PSK has excellent protection against noise because the information is contained within its phase. Noise mainly affects the amplitude of the carrier. One of the most popular PSK modulation methods used in LMDS is quadrature phase shift keying (QPSK). QPSK is easy to implement and fairly resistant to noise. It can be most easily considered as cosine and sine waves as given by the functions:

$$\phi_1(t) = \sqrt{2/T_s}\,\cos[2\pi f_c t]$$

$$\phi_2(t) = \sqrt{2/T}\,\sin[2\pi f_c t]$$

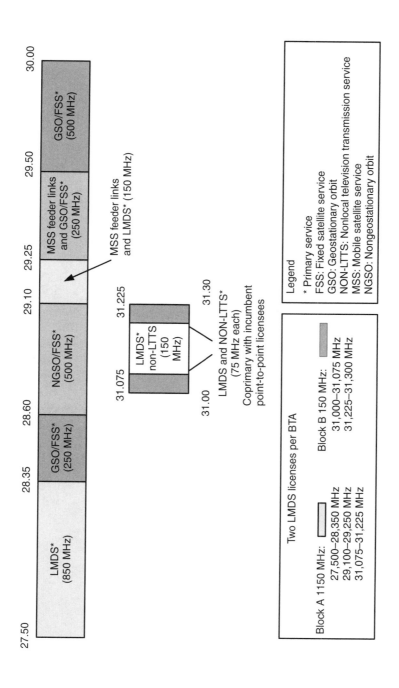

Figure 3.8 28- and 31-GHz LMDS band allocation in the United States.

T_s is the symbol duration, where each symbol represents two binary digits.

f_c is the carrier frequency. The coordinates of the signal constellation (or space) diagram are given by

$$\chi_1 = \int_0^{T_s} r(t)\phi_1(t)dt$$

$$\chi_2 = \int_0^{T_s} r(t)\phi_2(t)dt$$

where $r(t)$ is the received signal.

Each waveform can represent two binary digits. These functions are *orthogonal*, which roughly means they are at right angles. The more precise definition of orthogonality is that the product integrated over one cycle is zero, a property that a cosine and sine of the same frequency exhibit. Note *quadrature* also describes the 90 degrees of lag that sine maintains relative to a cosine.

The signal constellation (or space) diagram for QPSK is shown in Figure 3.9. Signal point B (11) and D (10) differ in only binary digit position. This type of mapping is called *gray encoding*. If, for example, signal point A is transmitted and a symbol error occurs, it is very likely that the received symbol will either be C or D. The gray encoding scheme used will mean that, on

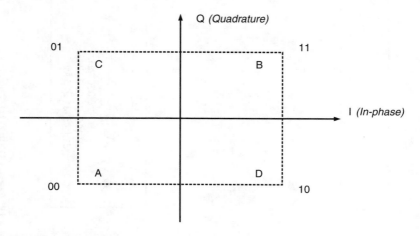

Figure 3.9 Signal constellation for QPSK.

average, we can expect the probability of an error in a binary digit to be half the probability of an error in a symbol.

The popular AM modulation is quadrature amplitude modulation (QAM). QAM is a modulation scheme in which two quadrature carriers, both a sine and a cosine wave, are amplitude modulated in accordance with a sequence of information bits. This comes in several flavors: 4 QAM, 16 QAM, 64 QAM, and higher. The higher the QAM, the more bits that can be transmitted in a hertz of spectrum. In 16 QAM, 4 bits of information are sent in one symbol interval. In 64 QAM, 6 bits of information are sent in the same interval. Numbers indicate the number of code points per symbol. The QAM rate or the number of points in the QAM constellation can be computed by 2^N bits per symbol. From a performance standpoint, this means that 64 QAM is 50% more bandwidth-efficient than 16 QAM, since we are sending more information in the same time interval. 16 QAM needs 16 distinct transmission signals, as shown in Figure 3.10.

While it is more spectrally efficient, the downside to using 64 QAM is its complexity. To send more bits per symbol, we need 64 distinct transmission signals instead of just 16, as shown in Figure 3.11. To keep the signals as different as possible, and to keep the symbol error rate the same, we must

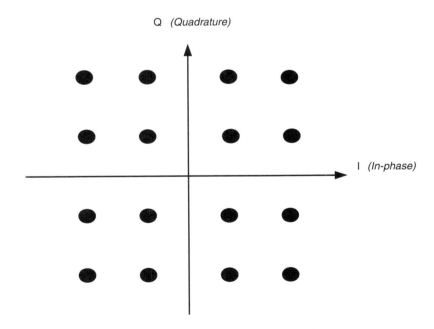

Figure 3.10 Signal constellation for 16 QAM.

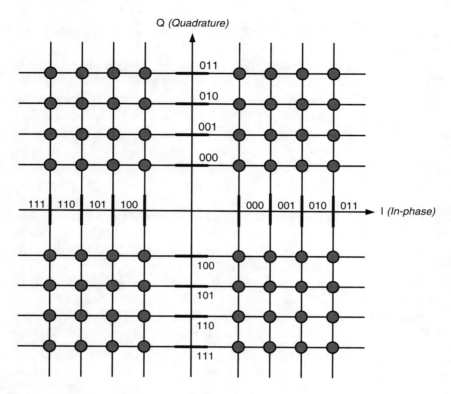

Figure 3.11 Signal constellation for 64 QAM.

have signals with greater amplitudes and, therefore, with more power. 64 QAM takes about 6 dB more transmitted power than 16 QAM for the same error performance. The price for more bits per hertz is the need for a much cleaner signal at the receiving site so that the more tightly packed bits can be recovered.

With a FCC limit on radiated power, this means reduced range. Thus, the trade-off becomes either lots of data-carrying capacity or longer range—which should not be a problem if the LMDS system supports multiple modulation methods in the same sector. Table 3.1 gives an estimated scale as to the amount of bandwidth these modulation schemes require for a 2-Mbps CBR connection [without accounting for overhead due to ATM and forward error correction (FEC)]. Spectral efficiency is measured in bits per second per hertz, a basic figure of merit for different modulation schemes. For example, for 16-QAM modulation, the spectral efficiency is 3.5 bps/Hz; for 64 QAM, it is 5 bps/Hz.

Table 3.1
Spectral Efficiencies

Name	Modulation Method	MHz for 2-Mbps CBR Connection (MHz)	Spectral Efficiency (bps/Hz)
QPSK	Quadrature phase shift keying	1.4	1.5
4 QAM	Quadrature amplitude modulation, 4 states	1.4	1.5
16 QAM	Quadrature amplitude modulation, 16 states	0.6	3.5
64 QAM	Quadrature amplitude modulation, 64 states	0.4	5.0

A digital modulator accepts a digital stream and provides a QPSK, 4-QAM, 16-QAM, or 64-QAM intermediate frequency (IF) signal delivery over the LMDS bandwidth. The modulator performs all the functions required to modulate digital video, voice, and data to a standard IF for input to the wireless transmitters. A QAM demodulator contains two separately addressable demodulator channels, each capable of accepting QPSK, 4-QAM, 16-QAM, and 64-QAM signals at symbol rates between 1 and 10 megasymbols per second (Msps).

Link budget is used to estimate the maximum distance that a subscriber can be located from a cell site while still achieving acceptable service reliability. The budget accounts for all system gains and losses through various types of equipment. The link budget analyzes several network parameters, including carrier-to-noise ratios (CNRs); carrier-to-composite triple beat ratios; and self-repeat site interference [carrier/interference (C/I)] and link-fade margins. In some cases, the microwave equipment is channelized to support a single carrier. Other systems offer broadband multichannel capability in which multiple carriers can be supported through a single transmitter.

3.4.5   Wireless Access Methods

In order to provide two-way (duplex) communications between LMDS base stations and remote sites, two methods known as frequency division duplex (FDD) and time division duplex (TDD) are used. FDD uses two separate frequencies, one for the uplink (remote site to base station) and one for the downlink (base station to remote site). TDD uses the same frequency for uplink and downlink communication. The base station and the remote take turns transmitting. Hence, the remote site and base station must agree on the timeslot for uplink and downlink transmission.

Multiple access methods, such as time division multiple access (TDMA), frequency division multiple access (FDMA), and code division multiple access (CDMA), are used in combination with FDD and TDD to increase the capacity of a wireless system. In FDMA, a block of frequency spectrum is divided into several distinct bands known as channels. Each channel occupies one of the distinct bands for the whole duration of transmission. In TDMA, each channel is divided into multiple timeslots, each of which supports an individual data burst. This technique is very similar to TDM in wireline T1/E1 transmission. The total available bandwidth, the bandwidth of the individual channels, and the number of timeslots per channel vary according to a particular standard, as well as the coding technique applied.

In the downlink direction, from the base station to customer premises, most LMDS carriers supply TDM streams, either to a specific remote site (point-to-point connectivity) or to multiple remote sites (a PMP system design). Figure 3.12 shows an FDMA scheme in which multiple remote sites share the downlink connection. Separate frequency allocations are used from each remote site to the base station.

Figure 3.13 shows a TDMA scheme in which multiple remote sites share both downlink and uplink channels. With FDMA and TDMA access links, whether downlink or uplink, there are a number of factors that affect efficiency and usage. For FDMA links, each customer site is allocated bandwidth that is either constant or varied over time. For TDMA links, the

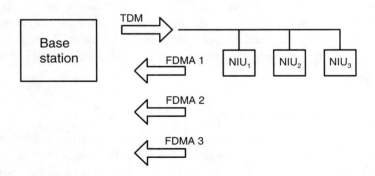

Figure 3.12 FDMA access scheme.

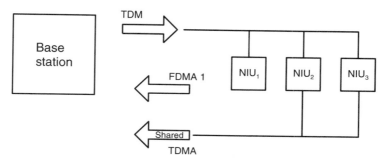

Figure 3.13 TDMA access scheme.

remote site is allocated bandwidth designed to respond to data bursts from the remote site.

There are several factors that may effect which system is best for certain applications. If the traffic is smooth enough (e.g., PCM voice), the uplink traffic requirements can be handled effectively using FDMA techniques. FDMA provides a constant pipe, with bursting occurring based on fairness algorithms within the customer premises multiplexer.

Alternatively, if burstiness persists within the traffic stream, TDMA may be a better choice. The air link protocol provides communication between the base station and the remote stations for the control and management of the air resources. In TDMA-based systems, the Medium Access Control (MAC) protocol allocates time slots to different remote sites. Each remote site can transmit only if it has been given a time slot. TDMA systems allow bursting to maximum rate of the channel, based on fairness algorithms implemented in the wireless MAC and the customer-premises multiplexer. TDMA allows for bursty response and does not request slots unless necessary. In other words, when one user is not using the spectrum, another one can use it. In FDMA systems, the link is always on, regardless of whether or not the user sends data.

In TDMA, MAC efficiency ranges from 65% to 90% or higher, depending on the burstiness characteristics of the users and the MAC design. In FDMA, efficiency is estimated at 100%, as there is no MAC. Both FDMA and TDMA systems allow high-priority user traffic to be sent first. Both systems multiplex various streams through the same wireless pipe. Channel efficiency is estimated at 88%, based on preamble and ranging in TDMA systems. In FDMA systems, the efficiency is 100%. However, open channel bandwidth may go unused in FDMA systems when one station has no data to send.

3.5 Backhaul Network

LMDS, implemented with a multiservice protocol such as ATM as its interface to the backhaul network, can transport voice, Internet, Ethernet, video, computer files, transaction data, and other traffic.

Standard network interfaces are provided at the ingress/egress of an ATM or PDH-based local-access platform for integration with backhaul networks. Hence, service providers can easily integrate different access platforms to rapidly deploy multiple services in their target markets.

Using ATM technology, the LMDS networks are able to provide voice, data, and video service to the buildings in the cell with a QoS guarantee using ATM Forum interworking and QoS concepts. Some LMDS products have IP-based services rather than ATM-based services. An IP architecture may be attractive for data-centric applications, but perhaps with more difficult QoS issues.

The LMDS radio provides the physical link. The choice of transport protocol is very important because it will determine the system's operational characteristics, as well as the types of services that can be offered to the user. The important characteristics are: support for all types of cargo (time sensitive such as real-time video and voice, as well as time-insensitive cargo such as e-mail); ability to withstand error bursts (system performance must be maintained); and support for standards that promote availability of low-cost silicon. ATM meets all of these requirements. It easily supports all types of data and even permits classes of service within data groups. All types of legacy traffic, as well as currently anticipated traffic, can be readily accommodated.

ATM's short (53-byte) packet size enables inexpensive, wireline speed-forward error correction and minimizes the throughput penalty that longer packet sizes impose. An error burst during a 64-KB packet transmission will necessitate the retransmission of all 64-KB. Within an ATM-based system, retransmission can occur in 53-byte increments.

Sections 3.5.1 and 3.5.2 provide an overview of the type of ATM-based backhaul network the wireless segment is connected to and the possible architectural changes that are likely to happen in the future.

3.5.1 Optical Networking Standards

Fiber-optic transmission systems that were deployed during the early stages were asynchronous in nature and often had to deal with high bit errors due to large variation in timing between transmitting and receiving ends. Also, the proprietary methods used by different manufacturers in formatting the optical signal gave rise to interconnection difficulties. These problems led to the development of ITU SDH (Recommendations G.707 [3], G.708 [4], and G.709 [5]) and ANSI SONET T1.105 standards.

The ITU SDH recommendations define a number of transmission rates, referred to as synchronous transport module (STM)–*n*, as shown in Table 3.2. STM-1 (155 Mbps) is the basic building block of carrying a number of standard low-rate signals as its payload. Similarly, SONET defines a basic building block called STS Level 1 (STS-1) (51.840 Mbps).

ITU and ANSI specifications define OC interfaces, as well as their electrical equivalents. Traditional photonic systems are, in practice, electro-opto-electronic (EOE) systems, as the origination and termination of a signal segment are electrically based. In this traditional architecture, the service layer (network switches, servers) interfaces with the SONET layer (transport and multiplexing functions). Add-drop, cross-connect, and restoration functions are performed in the electrical domain to support the TDM features of voice communication equipment.

3.5.2 Evolution of Optical Networks

The LMDS technology and its use within a multilayered architecture were reviewed in earlier sections in this chapter. Current backhaul networks, which wireless segments interface to, are predominantly ATM-based fiber-optic networks supporting comprehensive QoS guarantees. These networks are likely to evolve to optical networks and will also likely support IP and MPLS protocols. It is a formidable challenge to wireless network designers to provide seamless interworking arrangements and also support QoS guarantees when the end-to-end user session involves a wireless segment.

Table 3.2
Optical Transmission Rates

Frame Format	Optical	Bit Rate	Maximum DSOs
STS-1	OC-1	51.84 Mbps	672
STS-3 (STM-1)	OC-3	155.52 Mbps	2,016
STS-12 (STM-4)	OC-12	622.08 Mbps	8,064
STS-24 (STM-8)	OC-24	1.244 Gbps	16,128
STS-48 (STM-16)	OC-48	2.488 Gbps	32,256
STS-192 (STM-64)	OC-192	9.953 Gbps	129,024

Switching capacity limitations, bandwidth limitations, and the cost of electronic switches in this traditional architecture motivate carriers to migrate from the SONET layer to the optical layer.

Backbone networks are currently based on predominantly ATM or IP/PPP over SONET transport schemes at OC-3 to OC-192 speeds. There is increasing interest in operating these protocols directly over wavelength (λ) while still using the SONET framing. This scheme eliminates the need for costly SONET EOE gear and facilitates multiple λs over a single fiber commonly known as DWDM. Wavelength routers in such optical networks will switch inbound λ to an outbound λ. Soon, DWDM systems will be capable of supporting 160 to 240 wavelengths on a single fiber. The move to optical space will take place as the SONET layer functions, such as add/drop multiplexing, are reproduced to occur at the optical space [6, 7].

Multiprotocol lambda switching (MPLmS), which builds MPLS technology, is being defined by standard groups, such as the Optical Internetworking Forum (OIF), Optical Domain Service Interconnect (ODSI), and the IETF. MPLS IGP-TE extensions for optical networks and signaling extensions for RSVP and Label Distribution Protocol (LDP) are being worked out in these standard bodies.

References

[1]   Highsmith, W. R., "Converging Complementary Technologies To Maximize Service Flexibility," P-COM, Inc., *7th Annual Wireless Communications Association International Symposium,* Melbourne, FL, 1994, (Note: Section 3.1 is based primarily on information contained in this article).

[2]   McKenna, T., "Frame Relay's Heart Still Beats," *Telecommunications* March 2001, pp. 41–42.

[3]   International Telecommunications Union, "General Aspects of Digital Transmission Systems—Synchronous Digital Hierarchy Bit Rates," ITU-T Recommendation G.707, March 1993.

[4]   International Telecommunications Union, "General Aspects of Digital Transmission Systems—Network Node Interface for the Synchronous Digital Hierarchy," ITU-T Recommendation G.708, March 1993.

[5]   International Telecommunications Union, "General Aspects of Digital Transmission Systems—Synchronous Multiplexing Structure," ITU-T Recommendation G.709, March 1993.

[6]   Bala, K., "Making the New Public Network a Reality," *Optical Networks Magazine* May/June 2001, pp. 6–10.

[7]   Bala, K., "Internetworking Between the IP and the Optical Layer," *Optical Networks Magazine* May/June 2001, pp. 16–18.

4

User Access Methods That Need ATM Interworking

4.1 Introduction

This chapter describes user access methods that need ATM interworking support. The first method covered is multiprotocol encapsulation over Frame Relay. As described further in Chapter 6, Frame Relay service interworking (FRF.8) converts RFC 1490 [1] Frame Relay encapsulation into RFC 1483 [2] ATM encapsulation. This enables IP traffic (routed PDU traffic) carried by frames to be translated into ATM cells for service interworking.

RFC 1490 [1] encapsulation is widely used to transport IP over Frame Relay networks. RFC 1483 [2] encapsulation is used to carry IP traffic over ATM networks, as described further in Chapter 5. The similarities between these two methods of encapsulation are minimal, except for the encapsulation headers. Unfortunately, RFC 1490 [1] specifies a Network Layer Protocol ID (NLPID), subnetwork attachment point (SNAP) format for Frame Relay, while RFC 1483 [2] specifies a LAN-compatible LLC SNAP format. As a result, translation is required at the interworking function between Frame Relay and ATM SNAP formats, which are described further in Chapter 6.

This chapter also explains legacy protocols such as X.25 and SNA over Frame Relay, which again can be carried over ATM using RFC 1490 [1] to RFC 1483 [2] translation (FRF.8). The IP can also be used as an ATM network access protocol in cases where Data-Link Switching (DLSw) or Novell's IPX use TCP/IP for transport.

In addition, this chapter describes the physical access media used to transmit protocols over ATM. Digitized voice can be carried over DS1/E1 or DS3/E3 interfaces for the interworking that defines CES over ATM networks. Similarly, native mode ATM is carried over DS1/E1, DS3/E3, or STS-3c/STM-1 interfaces.

ATM may be used to carry all of the above-mentioned access protocols and has evolved into a truly broadband networking technology capable of carrying both voice and data applications with a given QoS. This chapter concludes by describing DSL, which provides high-speed Internet access using ATM encapsulation over modem technology and existing transmission media.

4.2 Multiprotocol Encapsulation over Frame Relay

This section describes encapsulation methods for carrying multiple protocols over Frame Relay networks. Frame Relay supports multiple user applications, such as TCP/IP, NetBIOS, SNA, and voice, thus eliminating the need for separate private line facilities to support each application. Because it statistically multiplexes traffic, Frame Relay allows multiple users at a location to access a single circuit and Frame Relay port, making more efficient use of the bandwidth.

Recognizing the ability of Frame Relay networks to carry multiple protocols, members of the IETF developed RFC 1490 [1], a standard method to encapsulate various protocols in Frame Relay. ANSI and the Frame Relay Forum enhanced this multiprotocol encapsulation method to include support for SNA protocols (FRF.3.1 [3]). CPE and terminal equipment that support multiprotocol encapsulation must know which Frame Relay virtual connections will carry which encapsulation method.

As described in Chapter 2, Frame Relay provides virtual connections between devices attached to the Frame Relay network, including point-to-point terminal connections and connections that are routed or bridged. Frame Relay virtual circuits are identified by DLCIs that have strictly local significance at each Frame Relay interface.

Encapsulation requires well-defined header formats that allow selection of the correct protocol stack at the remote end where encapsulated data will be processed. The Frame Relay network need not understand these header formats. They simply provide a transparent media for the transfer of data. This capability is provided using an identifier known as the Network Layer Protocol Identifier (NLPID) as defined in ISO/IEC TR 9577 and administered by ISO and CCITT, as shown in Figure 4.1. This field tells the receiver what encapsulation or protocol follows. For example, SNAP protocols are identified by a NLPID value of 0x80 and divided into bridged and routed categories. A NLPID value of 0x00 is defined within ISO/IEC TR 9577 as the null network layer.

All encapsulated protocols (e.g., IP, X.25, SNA) are preceded by an RFC 1490 [1] encapsulation header, as shown in Figure 4.1. The RFC 1490 [1] header contains information necessary to identify the protocol carried within the Frame Relay PDU, thus allowing the receiver to properly process the incoming packet.

The first octet of the RFC 1490 [1] header is the Q.922 [4] control field. For unacknowledged information transfer, the user information (UI) value 0x03 is used. For acknowledged HDLC elements of procedure (e.g., Q.922 [4]), information (I) and supervisory frames are used. I frames support layer 3 protocols, which require an acknowledged data-link layer such as the ISO 8208 protocols that form the data-link layer of X.25.

If a protocol can be encapsulated in more than one RFC 1490 [1] multiprotocol header format, the protocol is identified from one of the three methods below.

1. *Direct NLPID*—Used to identify protocols for which an NLPID value is defined in TR 9577 [e.g., IP, Connectionless Network Protocol (CLNP) (ISO 8473), and ISO 8208];

2. *SNAP encapsulation*—The NLPID value 0x80 followed by SNAP, as shown in Figure 4.2, used to carry LAN bridging and connectionless routed protocols;

3. *Other*—The four octets that follow NLPID 0x80 also used to identify layer 2 and layer 3 connection-oriented protocols (e.g., ISO 7776, Q.922 [4]) and protocols that cannot be supported by methods 1 and 2.

When using SNAP encapsulation, the SNAP header consists of an organizationally unique identifier (OUI) and a protocol identifier (PID). The three-octet OUI identifies the organization that administers the meaning of the following two-octet PID. For example, the OUI value of

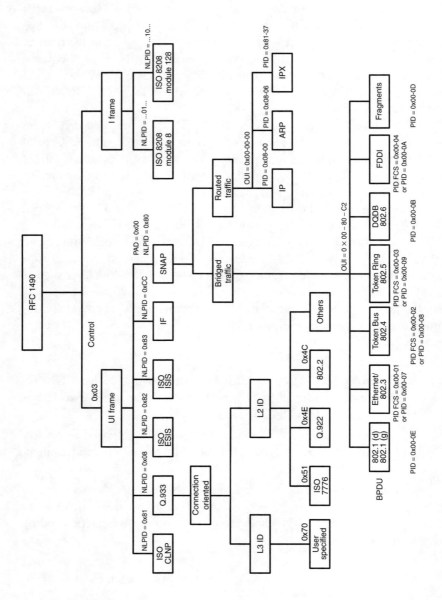

Figure 4.1 RFC 1490 multiprotocol encapsulation. (*From:* GNNettest, 1998.)

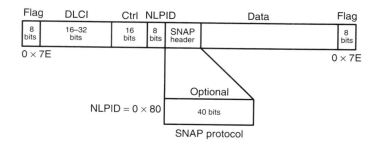

Figure 4.2 Format of the Frame Relay HDLC frame with RFC 1490 encapsulation.

0x0080C2 is the 802.1 [5] organization code. The PID that follows this value is administered by IEEE 802.1 [5]. The OUI and PID fields together identify a distinct routed, non-ISO routed, or bridged protocol. For example, an Ethertype value of 0x08-00 is an Internet IP PDU.

4.2.1 IP over Frame Relay

IP over Frame Relay allows LAN or client-server applications to be extended over distance-insensitive networks such as Frame Relay. The ease of configuring and changing virtual connections make Frame Relay ideal for meshed network configurations.

Packets are routed based on IP addressing when configured in the routed PDU mode, as shown in Figure 4.3. Bridged frames such as 802.3 [6] MAC frames are encapsulated and forwarded to the Frame Relay network in the bridged PDU mode, as shown in Figure 4.4. These interworking modes ensure interoperability with standards-based devices, including a wide variety of popular routers.

There are two ways of encapsulating IP PDUs. The first method uses a NLPID value of 0xCC, which is reserved for IP. The second method uses SNAP encapsulation. As described earlier, the type of routed PDU (layer 3) or bridged PDU (layer 2) is indicated in the SNAP header. When using SNAP encapsulation, the RFC 1490 [1] header will contain NLPID of 0x80, an OUI value of 0x00-00-00, and a PID value of 0x-08-00.

4.2.2 X.25 over Frame Relay

X.25 is another end-to-end service that uses RFC 1490 [1] encapsulation over a Frame Relay network. X.25 LAPB frames are encapsulated and sent across the Frame Relay network using Q.922 [4] control procedures, as

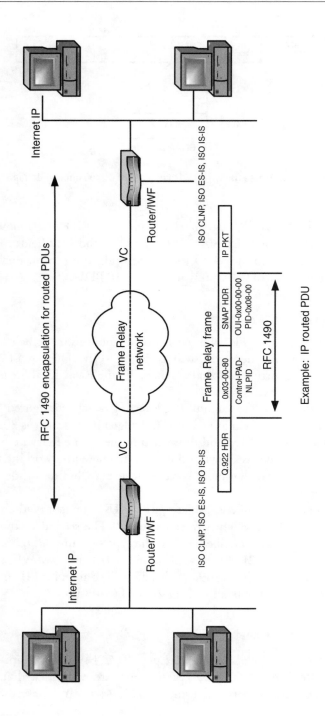

Figure 4.3 RFC 1490 encapsulation for routed PDUs.

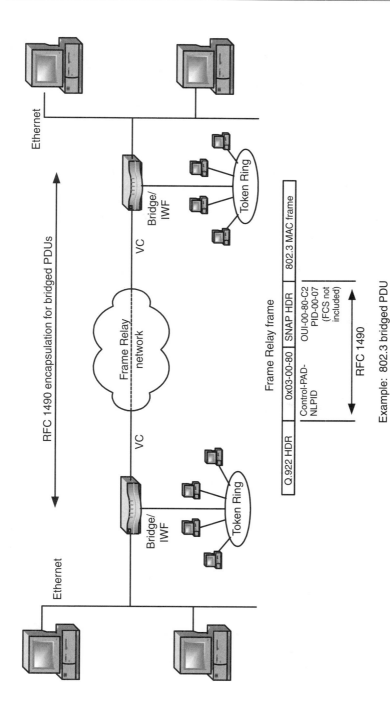

Figure 4.4 RFC 1490 encapsulation for bridged PDUs.

shown in Figure 4.5. The FRADs at either end of the Frame Relay network provide the interworking stacks that encapsulate and decapsulate the X.25 Link Access Protocol-Balanced (LAPB) mode frames, providing a LAPB logical link between the X.25 DTE and the X.25 packet network.

In this configuration, X.25 LAPB frames arriving at the FRAD from the X.25 DTE device are first encapsulated in accordance with RFC 1490 using NLPID values of 0x01 (modulo 8) or 0x10 (modulo 128). The Q.922 [4] Frame Relay packets that contain the RFC 1490 [1]/LAPB frames are carried on a configured Frame Relay PVC or SVC, as described in Chapter 2. The FRAD's interworking stack at the far end decapsulates the RFC 1490 [1] information and forwards the contained LAPB frames to the X.25 network, where the logical LAPB link is terminated at the DCE of the X.25 network. The data is delivered to the destination X.25 DTE via the X.25 network, where the X.25 VC terminates.

There may be many X.25 VCs multiplexed onto each Frame Relay connection, and they may terminate in many different locations in the X.25 network. Likewise, the X.25 DTE may interwork with multiple Frame Relay connections to reach many different destinations on the Frame Relay network.

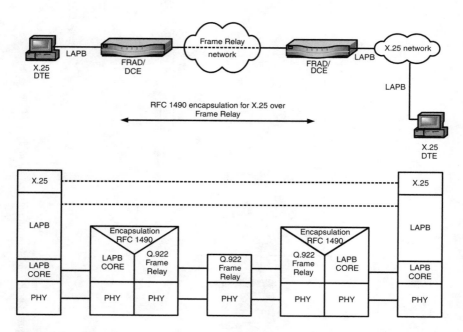

Figure 4.5 X.25 over Frame Relay with RFC 1490 encapsulation.

When sending LAPB frames across the Frame Relay network, the encapsulated LAPB address field, LAPB control field, and LAPB information field are sent transparently. The FCS in the LAPB frame is stripped off and a new FCS is calculated for the Q.922 [4] Frame Relay frame, including the Q.922 [4] address field (DLCI and FECN, BECN, DE, C/R, EA bits) and the Q.922 [4] information field. When the Q.922 [4] frame is entering a packet-switching network or device, the reverse procedure (decapsulation) occurs. The Q.922 [4] address field and FCS are stripped off, along with the RFC 1490 [1] information, and a new LAPB FCS is calculated.

These multiprotocol encapsulation procedures, as defined in RFC 1490 [1], provide one method of allowing interconnection of X.25 devices via a Frame Relay connection. Single-protocol X.25 encapsulation can also be used to provide interconnection of X.25 devices via Frame Relay.

4.2.3 SNA over Frame Relay

Frame Relay has emerged as a cost-effective and strategic WAN transport technology for SNA traffic. Frame Relay provides SNA customers with an unprecedented opportunity to reduce network costs dramatically while significantly increasing network bandwidth.

Traditional SNA networks are based on leased lines, which connect multiple controllers to each front-end processor (FEP). Frame Relay, as a virtual private-line replacement, offers straightforward migration from the complexities of multidrop leased lines. Migration can occur without any change to FEP hardware or software. Users can realize significantly lower monthly WAN costs, which can pay for a Frame Relay migration within months. Upgrading to Frame Relay allows fully meshed topologies for redundancy and backup without managing a large number of dedicated lines. Adding and deleting virtual connections can be done via network management and service subscription, instead of adding and deleting hardware.

There are two main approaches to sending SNA data over Frame Relay networks, namely RFC 1490, described in this section, and DLSw, described in Section 4.3.1. ANSI and the Frame Relay Forum enhanced the RFC 1490 [1] multiprotocol encapsulation method in FRF.3.1 [3], a standard that provides native mode support for SNA over Frame Relay, as shown in Figure 4.6. Native mode in this context means the ability to directly insert a complete SNA message unit within a Frame Relay frame without having either to encapsulate it within TCP/IP or to use some type of bridging scheme.

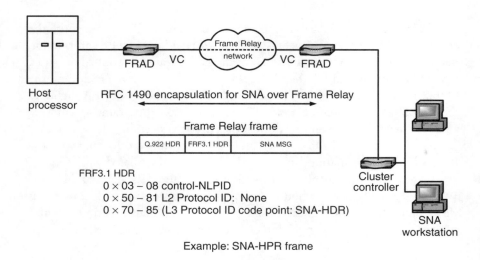

Example: SNA-HPR frame

Figure 4.6 SNA over Frame Relay with FRF3.1 encapsulation.

Typically, SNA controllers, routers, and FRADs encapsulate SNA as multiprotocol data for transport across Frame Relay networks. FRF.3.1 [3] specifies how to encapsulate SNA subarea, SNA/Advanced Peer-to-Peer Networking (APPN) [with and without high-performance routing (HPR)] within a RFC 1490 multiprotocol framework. Because encapsulated SNA data is transparent to the Frame Relay network, this allows multiple distinct protocols to be multiplexed across a single Frame Relay interface. Frame Relay access nodes are responsible for converting the user data into an FRF.3.1 [3] format appropriate for carrying SNA and LAN traffic.

Protocols that do not have a specific NLPID, such as SNA, use a NLPID value of 0x08. The four octets following the NLPID field identify both the layer 2 and layer 3 protocols being used. The code points for most protocols are defined in ITU-T Q.933. For example, the SNA HPR network layer packet (without layer 2) is identified by 0x50-0x81-0x70-0x85 (see Figure 4.6).

In addition, some RFC 1490 FRADs support different transmission strategies for SNA and LAN traffic. These FRADs ensure that SNA traffic is never sent out at speeds that will exceed the Frame Relay CIR, preventing this traffic from being arbitrarily discarded in the event of network congestion. Other non-SNA LAN traffic, which is much more tolerant of frame discards, is allowed to burst beyond the CIR for maximum performance.

4.3 IP As an Access Protocol

The popularity of TCP/IP has given rise to various types of interworking with this suite of protocols. The protocols that use IP as an access protocol are DLSw and Novell's IPX.

4.3.1 DLSw

DLSw provides a means of transporting IBM SNA and network basic input/output system (NetBIOS) traffic over an IP network. A data link switch typically supports IBM SNA and NetBIOS systems attached to IEEE 802.2–compliant [7] LANs, as well as SNA systems attached to IBM synchronous data-link control (SDLC) links. DLSw initially emerged as a proprietary IBM solution in 1992. DLSw is now documented in detail by IETF RFC 1795 [8], which was submitted in April 1995. It serves as an alternative to Source-Route Bridging (SRB), a protocol for transporting SNA and NetBIOS traffic in Token Ring environments that was widely deployed prior to the introduction of DLSw. In general, DLSw addresses some of the shortcomings of SRB for certain communication requirements, particularly in WAN implementations. RFC 1490 [1] interworking is also supported by DLSw, allowing LLC Type 2 over Frame Relay and DLSw prioritization.

The principal difference between DLSw and bridging is that for connection-oriented data, DLSw terminates the LLC connection at the DLSw switch. The DLSw multiplexes LLC connections onto a TCP connection to another DLSw. Therefore, the LLC connections at each end are totally independent of each other. Figures 4.7 and 4.8 show two end systems operating with bridged and switched LLC Type 2 services.

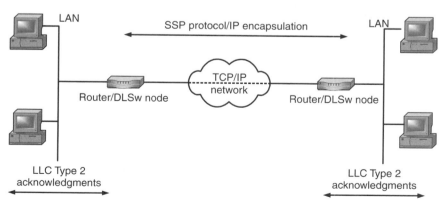

Figure 4.7 DLSw circuit over TCP/IP WAN.

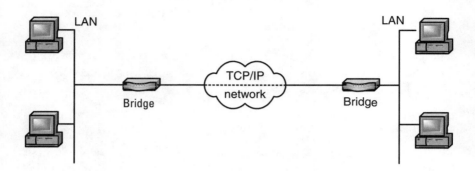

LLC Type 2 acknowledgements and IP encapsulation of information

Figure 4.8 LAN bridging over TCP/IP WAN.

IBM's SNA and NetBIOS are both classified as connection-oriented protocols. The DLC procedure that these protocols use on the LAN is LLC Type 2. NetBIOS also makes extensive use of datagram services that use connectionless LLC Type 1 service.

It is the responsibility of the DLSw to deliver frames that it has received from an LLC connection to the other end. TCP/IP is used between the DLSw to guarantee delivery of frames. As a result, LLC time-outs are limited to the local LAN (i.e., they do not traverse the wide area). Also, the LLC Type acknowledgments [receiver readies (RRs)] do not traverse the WAN, thereby reducing overhead across the wide-area links.

DLSw operation can be broken into the following three basic components:

1. *Capabilities exchange*—involves the trading of information about capabilities associated with a DLSw session;

2. *Circuit establishment*—occurs between end systems and includes locating the target end system and setting up DLC connections between each end system and its local router;

3. *Flow control*—enables the establishment of independent, unidirectional flow control between switches.

Three functions of the DLSw can be categorized as follows:

1. Support for the Switch-to-Switch Protocol (SSP), the peer-to-peer protocol maintained between two DLSw nodes or routers;

2. Termination of SNA DLC connection;

3. Local mapping of DLC connections to a DLSw circuit.

The SSP is used to establish connections, locate resources, forward data, and handle flow control and error recovery between DLSws. In general, SSP does not provide for full routing between nodes, because this is generally handled by common routing protocols such as RIP, OSPF, or Interior Gateway Routing Protocol (IGRP) or Enhanced IGRP (EIGRP). Instead, SSP switches packets at the SNA data-link layer. It also encapsulates packets in TCP/IP for transport over IP-based-networks and uses TCP as a means of reliable transport between DLSw nodes.

Native-mode RFC 1490 contrasts with DLSw, where all the SNA message units are first encapsulated within TCP/IP datagrams within 56-byte headers and then encapsulated into Frame Relay frames. Because DLSw was originally designed to transport SNA over router backbones, it includes substantial functionality to enable the reliable transport of connection-oriented traffic (SNA) over a connectionless (IP) network. The connection-oriented nature of Frame Relay, however, is inherently better suited to SNA. In an SNA environment, RFC 1490, which is designed for direct use of Frame Relay, provides more efficiency, reliability, and scalability, with less resource consumption.

4.3.2   IPX Packets over UDP/IP

Internet Packet Exchange (IPX) is a protocol developed by Novell to provide internetworking support for NetWare™ products. IPX is similar in functionality to IP.

IPX networking revolves around a scheme of numbered networks, unlike IP, which places more emphasis on network interface addresses. In an IPX environment, a network is a collection of equipment connected to the same LAN segment and using the same frame type. Different frame types on the same LAN segment are treated as separate networks. Each network must be allocated a number, which is unique across the entire network. IPX clients are given their network number by a server upon start-up, but they must know the correct frame type.

Routing between IPX networks is performed by adding two network cards in a server. This server then runs the RIP protocol to build a routing table for internetworking. Public broadcasts of this routing table are exchanged between servers. Within a short time, each server discovers the topology of the internetwork.

When there are two Novell LANs connected by IP, tunneling can be used for seamless connectivity. The IP tunnel provides a bridge-like facility by allowing IPX packets to be encapsulated within TCP/IP. Each IPX server listens for IPX packets, wraps each within a TCP/IP datagram, and routes the IP packet to a remote IP address. The remote IPX server strips off the TCP/IP wrapper and routes the IPX packet onto the destination IPX network.

4.4 Physical Media Interfaces to ATM Networks

This section examines various types of physical transmission media used for interfacing with ATM networks. The capacity of each transmission media (see Table 4.1) governs the bandwidth available to end users connected to CPE. For digitized voice, CES defines interworking with ATM. For data protocols, there are Frame Relay interworking (FRF.5 and FRF.8) and IP over ATM, to name two. There is also native-mode ATM, which simply transports ATM cells through the ATM network.

The CES supports 64K digitized voice channels using DS1/E1 or DS3/E3 as a transmission media. Frame Relay uses DS1/E1 transmission

Table 4.1
Physical Interfaces and Rates

	Interface	Rate
PDH	DS0	64 Kbps
PDH	DS1/T1 (DSX1)	1.536 Mbps (24 × DS0) + 8 Kbps (framing and signalling channel)
PDH	E1	2.048 Mbps (32 × DS0)
PDH	DS1c	2 × DS1
PDH	DS2	2 × DS1c
PDH	DS3/T3	7 × DS2, 28 × DS1/44.736 Mbps
PDH	E3	34.368 Mbps
SONET	STS-N (1–192)	STS-1/51.84 Mbps
	OC-N (1–192)	OC-3/155.52 Mbps
SDH	STM-M (1–64)	

media. IP over ATM may use Ethernet/Token Ring for the transmission of IP packets. Native ATM uses DS1/E1, DS3/E3, STS-*N* where *N* ranges from 1 to 192 and STM-*M*, where *M* ranges from 1 to 64, depending on the CPE and bandwidth needs. Thus, there is quite an array of transmission media that the interworking function has to deal with for each type of ATM interworking.

The Bell Labs (nearly synchronous) PDH was developed to carry digitized voice over twisted wire more efficiently in metropolitan areas. The lowest rate for digitized voice, called a digital stream 0 (DS0), is 64. The next rate is digital stream 1 (DS1), which bundles 24 DS0s onto a single twisted pair. A transmission repeater system over a four-wire twisted pair is called T1 or DS1. The DS1 transmission rate of 1.544 Mbps is achieved by multiplying the DS0 rate of 64 by 24 (i.e., 1.536 Mbps), plus an 8-Kbps framing and signaling channel.

International E1 standards apply a similar multiplexing technique to make better use of existing twisted pair, but multiplex 32 64-Kbps channels, totaling 2.048 Mbps. When it comes to data communications, only a discrete set of fixed rates are available, namely *n* × DS0, where *n* is 1 to 24 in North America (for DS1) or 1 to 31 in Europe (for E1).

The next higher rate is known as DS1c, which compromise two DS1s. Two DS1cs are multiplexed to give DS2. Finally, seven DS2s are multiplexed to give DS3 (T3) at the rate of 44.376 Mbps. As we can see, it takes 28 DS1s to achieve a rate of DS3 commonly used in inverse multiplexers (I-Muxes). ANSI standards refer to DS1 as DSX1 and DS3 as DSX3.

As the need for higher speed transmission grew with the Internet, optical fiber technology made it possible to achieve speeds greater than the standard DS3 44.376-Mbps rate. Two standards developed in the 1990s are the North American SONET and international SDH systems. These offer high transmission quality and direct multiplexing without the intermediate multiplexing stages of PDH.

The basic building block for SONET is the STS-1 at 51.84 Mbps. Higher transmission rates are defined in terms of STS-*N* that yield a rate of *N* × STS-1 (*N* × 51.84 Mbps), where *N* is from 1 to 192. OC-*N* refers to the optical characteristics of the signal that carries SONET payloads. It is a common practice for the terms STS-*N* and OC-*N* to be used interchangeably, but it is important to realize that an STS-*N* signal can only be carried on an OC-M, as long as *M* is greater or equal to *N*.

Similarly, the ITU developed the STM-1 with a rate of 155.52 Mbps to form the basic building block for optical transmission using the SDH standard. This rate is equal to SONET's STS-3 rate but is not interoperable due to differing overhead byte definitions. Similar to STS-N, higher rates are defined in terms of STM-M ($M \times 155.52$ Mbps), where M is from 1 to 64.

When transporting ATM cells through the network in native ATM mode, synchronization at the receiver is achieved by using the ATM HEC on every ATM cell. Thus, the HEC is used to locate cell boundaries in the received bit stream. Once the receiver locates several consecutive cell headers in the bit stream, then the receiver knows to expect the next cell 53 bytes later. Standards call this *HEC-based cell delineation*. Thus, the fixed length of ATM cells aids in detecting valid cells reliably. The ATM HEC is a 1-byte code applied to the 5-byte ATM cell header.

4.4.1 DS1/E1: CES

There is a user demand for carrying certain types of CBR, or circuit, traffic over ATM networks. As ATM is essentially a packet-oriented transmission technology, it must emulate circuit characteristics in order to provide good support for CBR traffic. The CES in ATM supports several types of interfaces such as DS1/E1 and DS3/E3. The service can be categorized into two basic service categories, namely unstructured and structured.

The unstructured DS1 service supports an interface speed of 1.544 Mbps. The unstructured E1 service supports an interface speed of 2.048 Mbps. These services provide a transparent transmission data stream across the ATM network, modeled after asynchronous DS1/E1 leased private lines.

The structured DS1/E1 service is defined in terms of $N \times 64$, where N is the number of time slots. The service can be configured to minimize ATM bandwidth by only sending the time slots that are actually needed to carry payload. The structured DS1/E1 can support CAS as an option, as described in Chapter 7.

4.4.2 DS1/E1: Frame Relay

Frame Relay is somewhat similar to a CES structured service when using a physical DS1/E1 interface without the CAS option. As the time slots are used to carry HDLC frames, there is no need for CAS. The Frame Relay data pipe (known as a logical port) can be configured in terms of $N \times 64$, where N is the number of time slots. Depending on the bandwidth requirements, the interface can be configured to a maximum of (24×64) 1.536 Mbps for DS1, and a maximum of (31×64) 1.984 Mbps for E1.

4.4.3   DS1/E1: Native ATM

Figure 4.9 shows direct cell delineation mapping for the DS1 physical interface. The bit rate for the DS1 interface is 1.544 Mbps. The physical layer interface format is the 24-frame multiframe extended superframe (ESF) for DS1 as defined in ANSI T1.403 [9]. ATM cells are carried in the DS1 payload (bits 2–193), in accordance with ITU G.804 [10], by using the direct mapping. ATM cell boundaries need not align with octet boundaries in DS1; ATM cells are byte-aligned to the DS1 frame.

Similarly, Figure 4.10 shows direct cell delineation mapping for the E1 physical interface. The bit rate for E1 is 2.048 Mbps, as specified in ITU-T Recommendation I.431 [11]. The E1 transmission frame consists of 32 time slots, numbered 0 to 31. Slots 0 (ITU-T Recommendation G.704 [12]) and 16 are reserved for framing, OAM, and signaling functions. Slots 1–15 and slots 17 to 31 are available for carrying data traffic (ATM cells). Slot 16 is reserved for signaling, as defined in ITU-T Recommendation G.804. The ATM cell is mapped into bits 9 to 128 and bits 137 to 256 (i.e., time slots 1 to 15 and time slots 17 to 31) of the 2.048-Mbps frame. It is important to note that there is no relationship between the beginning of an ATM cell and the beginning of a 2.048-Mbps transmission frame. The ATM cells can cross the E1 frame boundary.

F		ATM header cell 1		Payload 1	
F		Payload 1			
F		Payload 1	ATM header cell 1	Payload 2	
F		Payload 2			
F		Payload 2		ATM header cell 3	Payload 3
F		Payload 3			

←——————— ATM cell mapping field: 24 octets ———————→

Provides F3 OAM functions
   Detection of loss of frame alignment
   Performance monitoring (CRC-6)
   Transmission of RDI
   Performance reporting

Figure 4.9 DS1 direct cell delineation mapping.

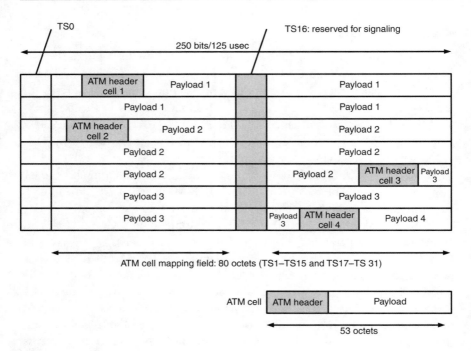

Figure 4.10   E1 direct cell delineation mapping.

On the transmit side, the DS1/E1 ATM UNI physical layer interface adapts the cell rate arriving from the ATM layer to the DS1/E1 frame payload capacity by inserting unassigned or idle cells when assigned cells are not available from the ATM layer. The receive side filters unassigned cells and idle cells, as described in ITU.432 [13], ANSI T1.627 [14], and ITU I.361 [15]. The cell delineation is performed using the HEC mechanism, as defined in ITU I.432 [13].

4.4.4   DS3/E3: Native ATM

Two different mappings are specified for transporting ATM cells over 44.736 Mbps DS3 interfaces, namely Physical Layer Convergence Protocol (PLCP) and a direct mapped system. These two mappings are not compatible, although both use the C-bit parity application of DS3 as the underlying transmission. C-bit parity is an application of the basic DS3 framing format, as described in ANSI T1.107 [16], "Digital Hierarchy—Formats Specifications" and ANSI T1.404 [17], "Network-to-Customer Installation—DS3 Metallic Interface Specification."

ATM cells can be directly mapped into the information payload, as shown in Figure 4.11 (transmission is from left to right); cells are nibble (nibble is 2 bits) aligned and inserted as consecutive bits into consecutive 84-bit info fields. Cells may start on any nibble boundary within any 84-bit info field and may cross *M*-frame boundaries. The ATM cell transfer capacity (for user information cells, signaling cells, OAM cells, and cells used for cell rate decoupling) is approximately 44.21 Mbps.

Carrying ATM traffic using DS3 PLCP is based on the IEEE 802.6 [18] Distributed Queue Dual Bus (DQDB) standard. Mapping of ATM cells into the DS3 PLCP is shown in Figure 4.12. The PLCP is then mapped into the DS3 information payload. Extraction of ATM cells from the DS3 operates in the analogous reverse procedure (i.e., by framing on the PLCP and then simply extracting the ATM cells directly). Because of the overhead induced by the PLCP, the nominal bit rate available for the transport of ATM cells in the DS3 PLCP is 40.704 Mbps.

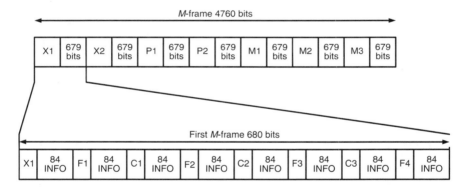

Overhead bit designation within *M*-subframes							
M-subframe 1	X1	F1	C1	F2	C2	F3	C3	F4
M-subframe 2	X2	F1	C1	F2	C2	F3	C3	F4
M-subframe 3	P1	F1	C1	F2	C2	F3	C3	F4
M-subframe 4	P2	F1	C1	F2	C2	F3	C3	F4
M-subframe 5	M1	F1	C1	F2	C2	F3	C3	F4
M-subframe 6	M2	F1	C1	F2	C2	F3	C3	F4
M-subframe 7	M3	F1	C1	F2	C2	F3	C3	F4

Notes:
1) See ANSI T1.107 for description of X1, X2, P1, and P2 bits.
2) The M-frame alignment signal shall be M1 = 0, M2 = 1, and M3 = 0
3) The M-subframe alignment signal shall be F1 = 1, F2 = 0, F3 = 0, and F4 = 1.
4) C1, C2, and C3 bits shall be assigned according to ANSI T1.404 for the C-bit parity application.

Figure 4.11 DS3 direct cell mapping.

PLCP framing		POI	POH	PLCP payload	
A1	A2	P11	Z6	First ATM cell	
A1	A2	P10	Z5	ATM cell	
A1	A2	P9	Z4	ATM cell	
A1	A2	P8	Z3	ATM cell	
A1	A2	P7	Z2	ATM cell	
A1	A2	P6	Z1	ATM cell	
A1	A2	P5	X	ATM cell	
A1	A2	P4	B1	ATM cell	
A1	A2	P3	G1	ATM cell	
A1	A2	P2	X	ATM cell	
A1	A2	P1	X	ATM cell	
A1	A2	P0	C1	Twelfth ATM cell	Trailer
1 octet	1 octet	1 octet	1 octet	53 octets	13 or 14 nibbles

Object of BIP-8 calculation

POI = path overhead indicator
POH = path overhead
BIP-8 = bit interleaved parity-8
X = unassigned—receiver required to ignore
A1 = 11110110
A2 = 00101000
P0-P11 = path overhead identifier
Z1-Z6 = growth octets
B1 = PLCP bit interleaved parity-8 (BIP-8)
G1 = PLCP path status
    = AAAAXXXX = FEBE B1 count
    = XXXXAXXX = RAI
C1 = cycle stuff counter
Trailer nibbles = 1100

Figure 4.12 DS3 PLCP framing (125 μs).

The DS3 PLCP consists of a 125-μs frame within a standard DS3 payload. PLCP then envelopes the ATM cells within this 125-μs frame, which itself resides inside the DS3 M-frame. There is no fixed relationship between the PLCP frame and the DS3 frame (i.e., the DS3 PLCP may begin anywhere inside the DS3 payload). The DS3 PLCP frame (see Figure 4.12) consists of 12 rows of ATM cells, each preceded by 4 octets of overhead.

Although the DS3 PLCP is not aligned to the DS3 framing bits, the octets in the DS3 PLCP frame are nibble aligned to the DS3 payload envelope. Nibble (4 bits) stuffing is required after the twelfth cell, to frequency justify the 125-μs PLCP frame. Nibbles begin after the control bits (F, X, P, C, or M) of the DS3 frame. The order of transmission of all PLCP bits shown in Figure 4.12 is from left to right and top to bottom. The figure represents the most significant bit (MSB) on the left and the least significant bit (LSB) on the right.

For new implementations, the direct mapping of ATM cells into the 44.21 Mbps DS3 payload, similar to that defined earlier for DS3, is preferred. However, the need to support existing implementations, as previously defined in ANSI T1.646 [19], "Broadband ISDN User-Network Interfaces—Rates and Formats Specifications" (i.e., those using PLCP-based mapping) must be taken into account as a subscription by network operators and manufacturers.

The physical E3 interface for ATM operates at 34.368 Mbps, which is based on ITU-T recommendations and ETSI specifications. The frame structure is shown in Figure 4.13. There are 7 bytes of overhead and 530 bytes of payload capacity per 125 μs. ATM cells are mapped into the 530 bytes of payload directly, as specified in Recommendation G.804 [12]. The bytes of the cells are aligned with the bytes of the frame. The mapping is shown in Figure 4.13.

4.4.5 STS-3c/STS-12: Native ATM

SONET is an international standard for high-speed optical communications. One of the physical layer interfaces that operates at bit rates of 155.52 Mbps is the interface based on the SONET STS-3c frame, the payload envelope necessary for the transport of ATM cells. The format of the STS-3c frame used at 155.52 Mbps is shown in Figure 4.14.

The mapping of ATM cells is performed by aligning, by row, the byte structure of every cell with the byte structure of the SONET STS-3c payload SPE. The entire STS-3c payload is filled with cells, yielding a transfer capacity of ATM cells of 149.760 Mbps (8,000 $\times$ 8 $\times$ 9 $\times$ 260). Because the STS-3c payload capacity is not an integer multiple of the cell length, a cell may cross an SPE boundary.

The user data rate is computed as 9 rows by 260 columns of bytes at 8,000 SONET frames per second, or 149.76 Mbps. The mapping of STS-12c is similar in nature. The difference between the North American SONET format and the international SDH format exists only in the TDM

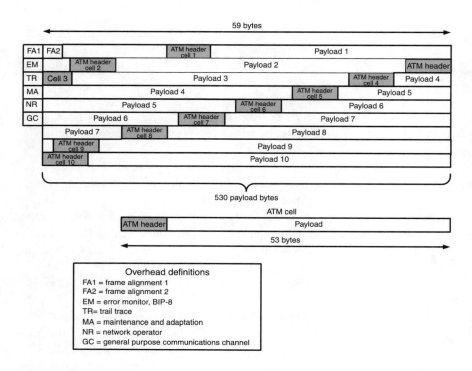

Figure 4.13 E3 direct cell mapping.

overhead bytes. The identification of cell boundaries in the payload for both formats is accomplished by the use of the HEC field in the cell header.

4.5 Broadband Services over DSL

Telecommunications companies are meeting demands for high-speed remote access to the Internet and corporate networks with the deployment of DSL technology. DSL provides a high-speed connection that offers unlimited opportunities for information on demand while facilitating improved on-line shopping, video gaming, and telecommuting experiences, to name a few.

DSL is often written as *x*DSL, indicating that it is a (growing) family of related standards and technologies, all designed to provide high-speed data communications over long spans of twisted-pair wire. The *x* stands for H, S, I, V, A, or RA, depending on the type of service offered [e.g., high-speed DSL (HDSL), symmetric or single-line DSL (SDSL)]. The types of DSLs

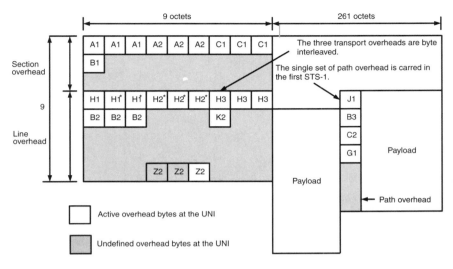

Overhead bytes: A1, A2, C1, H1, H2, H3, B2 are replicated
Overhead bytes: B1, K2, Z2 are not replicated

*Bits of H1, H2 bytes are set to all 1s for path AIS

```
A1, A2 = framing
C1 = STS-1 1D (1,2,3)
B1 = section BIP-8
H1 = (bits 1-4) = new data flag, path AIS
H1* = Concatenation Indicator, path AIS
H1, H2 (bits 7-16) = pointer value, path AIS
H2* = concatenation indicator, path AIS
H3 = pointer action, path AIS
B2 = Line BIP-24
K2 = Line AIS, RDI
Third Z2 byte = line FEBE
J1 = path Trace
B3 = path BIP-8
C2 = path signal label
G1 = path FEBE, RAI, FERF
Undefined overhead bytes at the UNI
```

Figure 4.14 SONET STS-3C format.

and corresponding rates are shown in Table 4.2. We will first provide an overview of these different technologies and then discuss an example of DSL connectivity with ATM.

At its simplest, DSL technology is just next-generation modem technology, as shown in Figure 4.15. It takes advantage of the fact that the loop—the copper pair connecting a home or office to the telecommunications company's central office—is a reasonably wide band medium, and all band limiting that holds conventional modems to 56 Kbps or less occurs in the central office or core network. The biggest difference between DSL and

Table 4.2
DSL Versions and Rates

DSL Version	Rate	Range
High-bit rate DSL (HDSL)	1.5 Mbps full duplex	12,000 ft
Symmetric DSL (SDSL)	160 Kbps–2.048 Mbps (E1)	1.3–3.1 mi
Asymmetric DSL (ADSL)	High-speed downstream 1.5–8 Mbps	5 mi
	Medium-speed duplex channel 16–640 Kbps	
ISDN DSL (IDSL)	144 Kbps	3.4 mi
Very-high bit rate DSL (VDSL)	13 Mbps downstream 1.6 Mbps upstream	1,500–4,500 ft
	Maximum 26 Mbps downstream 3.2 Mbps upstream	
	Speed varies with distance	

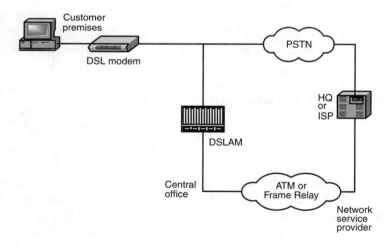

Figure 4.15 DSL wiring. (*From:* Global Knowledge.)

analog is not in technology or data rates but in application. Analog modems are always physically located at the origination and destination of the analog data traffic (i.e., a subscriber and an ISP facility). With DSL, one of the modems must be located in the telephone company's central office. It becomes the telecommunications company's responsibility to recover the subscriber's data from analog and transfer it to a pure digital network for delivery to the destination, perhaps via another DSL link. This function is performed by a DSL access multiplexer (DSLAM).

DSL implementations can be divided into two broad categories, depending on the method they use to place the data on the twisted pairs: baseband or passband. ISDN, ISDN DSL (IDSL), and HDSL use a baseband approach and asymmetric DSL (ADSL) uses a passband approach. Baseband systems have a frequency spectrum that extends down to zero frequency, while passband systems have a spectrum whose lower limit is at a nonzero frequency, potentially letting other traffic use lower frequencies. Voice service requires the 0–4-kHz portion of the spectrum, which is why systems like HDSL cannot coexist with voice service on the same twisted pair. Passband systems generate two or more channels, well above the baseband that contains amplitude and phase-modulated signals similar to those used by analog modems. Since all the data traffic is carried in these high-frequency channels, the baseband portions of the spectrum is free to support voice service. Passband systems are the preferred choice for our personal broadband service, since one of its goals is to preserve the 0–4-kHz voice channel. But how did this technology manage to deliver 1.5 Mbps down a highly lossy, impaired copper loop?

The spectrum above voice frequencies is divided into as many as 256 very narrow channels, called bins. Each of these channels is 4-kHz wide. Again, amplitude and phase modulation is used to place data into each channel. The overall effect is as if the data to be transmitted were divided into 247 separate streams, each were stream-fed to a modem, and the modems were stacked in frequency. The voice band is well spaced from the data channels, and a simple POTS splitter can separate voice from ADSL. The data spectrum ranges from about 32 kHz to a little over 1 MHz.

Figure 4.16 shows various types of DSL that are currently deployed. HDSL, the first version of DSL introduced, provides a full duplex DS1 over twisted pair up to 12,000-feet long. Developed by Bellcore in the late 1980s, it was intended to be an economical method of satisfying the exploding corporate demand for DS1 services. HDSL requires two unidirectional pairs to provide a symmetric 1.5 Mbps service but cannot coexist with voice telephone services on the same pair. Supported by several equipment vendors, HDSL is a well-defined technical standard and is widely deployed in networks around the world.

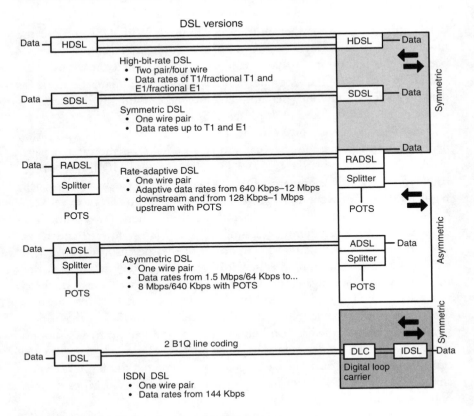

Figure 4.16 DSL versions. (*From:* Global Knowledge.)

SDSL is distinct from HDSL in that it does operate over a single twisted pair. In addition, SDSL allows transport of normal voice service over the same pair. SDSL is being offered over a variety of data rates, ranging from 160 Kbps up to an E1 at 2.048 Mbps. SDSL is a narrowly targeted service, available only where a user is close to the central office, twisted pairs are scarce, and the upstream bandwidth is as important as downstream.

ADSL works over conditioned telephone lines that transmit more than 6 Mbps to a subscriber, and as much as 640 Kbps in both directions. An ADSL circuit connects an ADSL modem on each end of a single twisted-pair telephone line, creating three information channels—a high-speed downstream channel, a medium-speed duplex channel (depending on the implementation of the ADSL architecture), and a POTS or ISDN channel. The high-speed channel ranges from 1.5 to 8 Mbps, while the duplex rates range from 16 to 640 Kbps. ADSL modems will accommodate ATM transport with variable rates and IP protocols.

Rate adaptive DSL (RADSL) is nothing more than an "intelligent" version of ADSL. Just as analog modems can "train down" to the rate necessary to establish a reliable connection, RADSL modems can automatically assess the condition of the twisted pair connecting them and optimize the line rate to accommodate the perceived line quality. RADSL allows the service provider to provision DSL service without having to measure a line and manually adjust or choose a modem to match. IDSL functions on the basic rate interface (BRI) model of ISDN, which provides an overall data transmission rate of 144 Kbps. The two bearer (B) channels are circuit-switched and carry 64 Kbps of either voice or data in either direction. The data (D) channel carries control signals and customer call control data in a packet-switched mode, operating at 16 Kbps.

Very-high-bit-rate DSL (VDSL) has the option of either symmetric or asymmetric transmission. The trade-off for high transmission speed is its distance (see Table 4.2). VDSL allows for the coexistence of digital and POTS transmission on the same twisted pair by using a POTS splitter.

A typical application of using ATM encapsulation over DSL is where IP packets are encapsulated using RFC 1483 multiprotocol at the IAD and modulated onto a DSL channel, as shown in Figure 4.17. IADs at the customer premises are key, in order to take advantage of the DSL infrastructure. Carrier class analog interfaces, coupled with a serial data port and integrated router (LAN), supply the interfaces needed to carry voice, data, and video services over a single WAN connection. The DSL network is typically ATM-based, and the DSLAM performs statistical multiplexing. The DSLAM provides connectivity to the ATM network. On the other end is an ATM router connected to the ATM network, which decapsulates the RFC 1483 packets and forwards the packets as IP to the LAN. ATM VP/VC

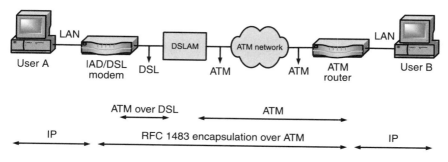

Figure 4.17 ATM encapsulation over DSL.

connection between the DSLAM and the ATM router provides the QoS for the connection.

ATM, with its signaling protocols, also provides the ability to establish direct, high-speed connections to other users or corporate locations as simply as making a phone call. While e-mail and Web-based applications and other IP-based connections will probably make up the bulk of our personal broadband service utilization, the ability to bypass the Internet altogether and establish more secure, direct, point-to-point connections with vendors, customers, or headquarters is a critical, positive differentiator that makes ATM over DSL attractive.

The final advantage of ATM over other options is that, in the long term, it is where networking is headed. Increasingly, the core of the Internet runs on ATM equipment, and ATM will gradually work its way out to the network edges.

References

[1] Bradley, T., C. Brown, and A. Malis, "Multiprotocol Interconnect over Frame Relay," RFC 1490, July 1993.

[2] Heinanen, H., "Multiprotocol Encapsulation over ATM Adaptation Layer-5," RFC 1483, July 1993.

[3] Frame Relay Forum Implementation Agreement FRF.3.1, "Multiprotocol Encapsulation Implementation Agreement (MEI)," June 1995.

[4] ITU-T Recommendation Q.922, "ISDN Data Link Layer Specification for Frame Mode Bearer Services," February 1992.

[5] IEEE Std 802—Overview and Architecture, May 31, 1990 (Note: This standard provides an overview to the family of IEEE 802 standards. This document forms part of the 802.1 scope of work.)

[6] ISO 8802-3 [ANSI/IEEE Std 802.3-1998], a bus using CSMA/CD as the access method.

[7] ISO/IEC 8802-2 [ANSI/IEEE Std 802.2-1998], Logical Link Control.

[8] Wells, L., and A. Bartky, "Data Link Switching: Switch-to-Switch Protocol AIW DLSw RIG: DLSw RIG: DLSw Closed Pages, DLSw Standard Version 1.0," RFC 1795, April 1999.

[9] ANSI T1.403.01, "Network and Customer Installation Interfaces—ISDN Primary Rate Layer 1 Metallic Interface Specification," 1999.

[10] ITU-T Recommendation I.431, "Primary Rate User-Network Interface—Layer 1 Specification," March 1993.

[11] ITU-T Recommendation G.704, "Synchronous Frame Structures Used at 1,544, 6,312, 2,048 and 44,736 Kbps hierarchical levels,"

[12] ITU-T Recommendation G.804, "ATM Cell Mapping into Plesiochronous Digital Hierarchy (PDH)," November 1993.

[13] ITU Recommendation I.432, "B-ISDN User Network Interface—Physical Layer Specification," March 1993.

[14] ANSI T1.627, "B-ISDN-ATM Layer Functionality and Specification," 1993 (Revised in 1999).

[15] ITU-T Recommendation I.361, "B-ISDN ATM Layer Specification," February 1999.

[16] ANSI T1.107, "Digital Hierarchy—Formats Specification," 2002.

[17] ANSI T1.404, "Network-to-Customer Installation—DS3 Metallic Interface Specification," 1994.

[18] ISO/IEC 8802-6 [ANSI/IEEE Std 802.6], using distributed queuing dual bus as the access method.

[19] ANSI T1.646, "Broadband ISDN—Physical Layer Specification for User-Network Interfaces, Including DS1/ATM," 1995.

Selected Bibliography

ATM Forum Technical Committee, "DS1 Physical Layer Specification," af-phy-0016.000, September 1994.

ATM Forum Technical Committee, "E1 Physical Interface Specification," af-phy-0064.000, September 1996.

ATM Forum Technical Committee, "ATM on Fractional E1/T1," af-phy-0130.00, October 1999.

ATM Forum Technical Committee, "DS3 Physical Layer Interface Specification," af-phy-0054.000, January 1996.

ATM Forum Technical Committee, "E3 Public UNI," af-phy-0034.000, August 1995.

ATM Forum Technical Committee, "Circuit Emulation Service Interoperability Specification Version 2.0," af-vtoa-0078.000, January 1997.

ATM Forum Technical Committee, "ATM User-Network Interface Specification Version 3.1."

Lane, J., Virata, "Personal Broadband Services: DSL and ATM," White Paper, Virata, September 1998.

5

ATM Interworking to Support Internetworking

5.1 Introduction

This chapter describes interworking of existing protocols such as IP and SS7 over ATM networks. The success of ATM depends on the ability to allow for interoperability between ATM and existing network and link- layer protocols in LANs and WANs, which currently have a large installed base. There are many different ways of running network-layer protocols from end to end, using ATM as a backbone protocol. Each method of interworking has its own merits, and the efficiency in using precious bandwidth becomes a major factor in their implementation.

The chapter begins with the multiprotocol encapsulation methods used with ATM interworking. The middle sections describe several interworking implementations of the popular IP with ATM. The classical IP over ATM implementation is described in detail, followed by the LANE protocol used to carry network-layer IP packets over ATM. Section 5.6 describes ATM interworking with SS7.

5.2 Multiprotocol Encapsulation over ATM

ATM-based networks are of increasing interest for both local- and wide-area applications. As a result, LAN- and WAN-based traffic will most likely be carried over ATM networks. This means that the different types of protocols commonly used in LANs and WANs will be carried over ATM backbone networks. Sometimes, it is economical to carry multiple protocols over a single ATM VP/VC as opposed to each protocol being carried by individual ATM VP/VCs.

RFC 1483 [1] multiprotocol encapsulation describes two different methods for carrying connectionless network interconnect traffic—routed and bridged PDUs—over an ATM network.

The first method, known as LLC encapsulation, allows several protocols to be carried over the same ATM VP/VC. Traffic is encapsulated with a header at the interworking function (IWF), as shown in Figure 5.1. A header is necessary to identify the protocol of the routed or bridged PDU. Packets are routed to the ATM network based on IP addressing when configured in the routed PDU mode. In the bridged PDU mode, bridged frames such as 802.3 [2] MAC frames are encapsulated and forwarded to the ATM network. LLC encapsulation ensures interoperability with standards-based

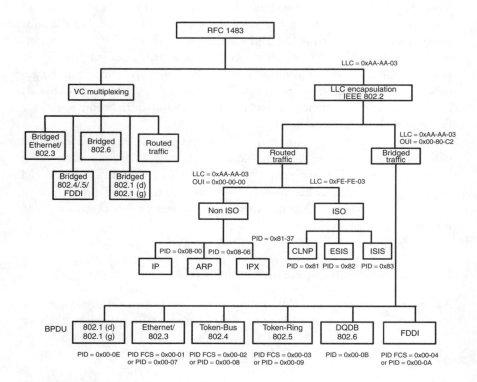

Figure 5.1 RFC 1483 multiprotocol encapsulation.

devices, including a wide variety of popular routers from such companies as Cisco, Bay Networks, ACC, 3 COM, Telebit, and IDEA.

The second method defined in RFC 1483 is VC-based multiplexing, which does not use LLC encapsulation. In this case, the routed or bridged PDUs are simply transported through the ATM network to their final destinations without encapsulation.

Either multiplexing method, LLC encapsulation or VC-based multiplexing, can be used with ATM PVCs and SVCs. The method is selected by a configuration option for PVCs. For SVCs, information elements in the ATM signaling protocol are used by the two interworking routers to communicate whether to employ LLC encapsulation or VC-based multiplexing. The LLC encapsulation method is the most commonly used when running IP over ATM, as described in Section 5.3. The ITU-T and ATM Forum have adopted this method as the default encapsulation for multiprotocol transport over ATM, as described in Section 5.5.

For IP-based applications, the method that is used for encapsulation can have an impact on efficiency. For example, an IP packet with a payload of 40 bytes, together with the AAL5 8-octet trailer, will fit exactly into a 48-byte ATM cell. If the LLC encapsulation method is used, which then adds an additional overhead of 8 bytes for the LLC header, it will generate an additional ATM cell. Thus, bandwidth efficiency on the ATM network will be nearly halved (58%) due to the LLC encapsulation overhead. Because it is estimated that almost one-third of IP packets are exactly 40-bytes long (TCP/IP acknowledgments), using VC-based multiplexing instead of LLC encapsulation can save considerable amount of ATM bandwidth. The following subsections describe these two methods in further detail.

5.2.1 RFC 1483 LLC Encapsulation

In RFC 1483 LLC encapsulation, the type of protocol is encoded in an IEEE 802.2 [3] LLC header placed in front of the carried PDU. The type of routed PDU (layer 3) or bridged PDU (layer 2) is also indicated in the LLC header.

The Institute of Electrical and Electronics Engineers (IEEE) subdivides the data-link layer into two sublayers: LLC and MAC. Communications via LANs use special functions of the MAC and LLC sublayers. The LLC sublayer, defined by 802.2, manages communications between devices over a single link of a network for both connectionless and connection-oriented services. In order to enable multiple higher layer protocols to share a single physical data link, the IEEE 802.2 LLC header consists of three one-octet fields: the destination service access point (DSAP), the source service access point (SSAP), and a control field.

The MAC sublayer manages protocol access to the physical network medium. The MAC protocols form the basis of LAN and MAN standards used by the IEEE 802.X LAN protocol suite, which includes Ethernet, Token Ring, and token bus.

The control field of the LLC header is always set to 0x03 for routed PDUs specifying that the type of PDU is an unnumbered information command PDU. For example, the LLC header value of 0xFE-FE-03 identifies that a routed ISO PDU follows (see Figure 5.2). The type of routed ISO-PDU is then identified by the NLPID field, which may be set to 0x81 (CLNP PDU), 0x82 [end system–to–intermediate system (ESIS) PDU] or 0x83 [intermediate system–to–intermediate system (ISIS) PDU].

In other cases, the LLC header value of 0xAA-AA-03 indicates the presence of an IEEE 802.1 [4] SNAP header. The SNAP header consists of an OUI and a PID. The three-octet OUI identifies the organization that administers the meaning of the following two-octet PID. For example, the OUI value of 0x0080C2 is the IEEE 802.1 [4] organization code. The OUI and PID fields together identify a distinct routed non-ISO routed or bridged protocol. For example, as shown in Figure 5.3, an OUI value of 0x00-00-00 specifies that the following PID is an Ethertype 0x08-00, which is an Internet IP PDU.

Similarly, bridged protocols are identified by the type of bridged media in the SNAP header, as shown in Figure 5.4. As with routed non-ISO PDUs, the presence of a SNAP header is indicated by the LLC header value 0xAA-AA-03. In addition, the PID indicates whether the original FCS is preserved within the bridged PDU.

Figure 5.5 illustrates protocol encapsulation by two routers, each multiplexing separate routed non-ISO and ISO PDUs over a single ATM

	IEEE 802.2 LLC header			
	DSAP	SSAP	CTRL	NLPID
CLNP PDU	FE	FE	03	81
ESIS PDU	FE	FE	03	82
ISIS PDU	FE	FE	03	83

Figure 5.2 LLC encapsulation of routed ISO-PDUs.

	IEEE 802.2 LLC header			SNAP header		
	DSAP	SSAP	CTRL	OUI	PID	PAD
IP	AA	AA	03	000000	0800	0000
ARP	AA	AA	03	000000	0806	0000
IPX	AA	AA	03	000000	8137	0000

Figure 5.3 LLC encapsulation of routed non-ISO-PDUs.

	IEEE 802.2 LLC header			SNAP header			
	DSAP	SSAP	CTRL	OUI	PID	PAD	
802.3 Ethernet	AA	AA	03	0080C2	0001	0000	LAN FCS
	AA	AA	03	0080C2	0007	0000	No LAN FCS
802.4 Token Bus	AA	AA	03	0080C2	0002	000000	LAN FCS
	AA	AA	03	0080C2	0008	000000	No LAN FCS
802.5 Token Ring	AA	AA	03	0080C2	0003	0000XX	LAN FCS
	AA	AA	03	0080C2	0009	0000XX	No LAN FCS
FDDI	AA	AA	03	0080C2	0004	000000	LAN FCS
	AA	AA	03	0080C2	000A	000000	No LAN FCS
802.6 DQDB	AA	AA	03	0080C2	000B		
BPDU	AA	AA	03	0080C2	000E		

Figure 5.4 LLC encapsulation of bridged PDUs.

VP/VC connecting the two locations. The routers multiplex the PDUs onto the same ATM VP/VC using the LLC encapsulations described earlier. Figure 5.6 illustrates how bridges multiplex the Ethernet and Token-Ring PDUs from an Ethernet and Token-Ring interface to yield a bridged LAN.

LLC encapsulation was initially designed for networks where devices would send all protocols over a single ATM VP/VC connection. Because

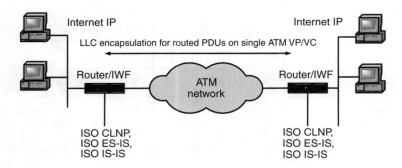

Figure 5.5 LLC encapsulation for routed PDUs.

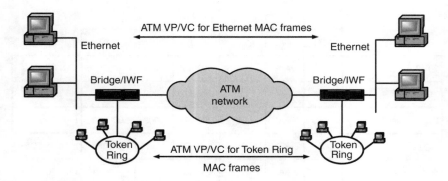

Figure 5.6 LLC encapsulation for bridged PDUs.

separate ATM VP/VC connections are not required for each protocol, the carrier can save money. The drawback of this method is that all packets get the same ATM QoS and efficiency is impacted by the overhead of the LLC/SNAP headers.

5.2.2 RFC 1483 VC-Based Multiplexing

In the second method of multiplexing specified by RFC 1483 [1], known as VC-based multiplexing, the carried network interconnect protocol is identified implicitly by the VP/VC connecting the two ATM end users. There is no need to include explicit multiplexing information (i.e., LLC header and SNAP header) in the payload of the AAL5 CPCS-PDU. This method minimizes bandwidth use and processing overhead at the expense of requiring separate VP/VC connections for each type of encapsulated protocol. RFC 1483 LLC encapsulation is more commonly used as the default method for

ATM interworking. This is the only method specified for Frame Relay (FRF.8) interworking with ATM, which is described in Chapter 6.

Figure 5.7 illustrates VC-based multiplexing of routed protocols, showing a separate ATM VP/VC connecting the routers at each end of the ATM network.

Figure 5.8 illustrates a similar configuration for bridged protocols, again requiring twice as many ATM VP/VCs as for LLC encapsulation.

5.3   IP Interworking with ATM

Interworking of IP with ATM networks is essential for ATM deployment because the IP is here to stay. There are different ways of running IP at the network layer while using ATM as a backbone protocol. One way is to use

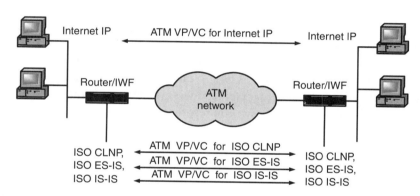

Figure 5.7 VC-based multiplexing for routed PDUs.

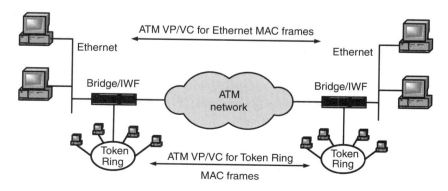

Figure 5.8 VC-based multiplexing for bridged PDUs.

address resolution to map network-layer addresses directly into ATM addresses, such as in classical IP over ATM. Another method is to carry network-layer IP packets across an ATM network by emulating a LAN on top of the ATM network. The following subsections describe these two alternatives and some of the issues associated with them.

5.3.1   Classical IP over ATM

RFC 1577/2225 [5] defines classical IP (clip) transport over ATM. The term classical indicates that the ATM network has the same properties as existing legacy LANs, connected through IP routers. In order to operate IP over ATM, a mechanism must be used to resolve IP addresses to their corresponding ATM addresses.

Clip defines two basic concepts: packet encapsulation and address resolution. Reusing the same connection for all data transfers between end users conserves connection resource space and saves on connection setup latency, after the first connection. To enable connection reuse, the packet must be prefixed with a multiplexing field. We discussed the two methods for doing this in Section 5.2. Clip over ATM is different from IP in legacy LANs in that ATM provides a virtual connection environment through the use of PVCs and/or SVCs. Once a clip connection has been established, IP datagrams are most often encapsulated using IEEE 802.2 [3] LLC/SNAP and segmented into ATM cells using ATM adaptation layer type 5 (AAL5).

Figure 5.9 illustrates three routers connected across an ATM network. If one router receives a packet across a LAN interface, it will first check its next-hop table to determine through which port, and to what next-hop router, it should forward the packet. If this lookup indicates that the packet is to be sent across an ATM interface, the router will then need to consult an address resolution table to determine the ATM address of the destination next-hop router. The table could also provide the ATM VP/VC value of a PVC connecting the two routers. The address resolution table could be configured manually, but this is not a very scalable solution. RFC 1577/2225 [5] defines a protocol to support automatic address resolution of IP addresses. The classical IP over ATM model introduces the concept of a logical IP subnet (LIS). Like a normal IP subnet, a LIS consists of a group of IP nodes (e.g., hosts and routers) that connect to a single ATM network and belong to the same IP subnet.

To resolve the addresses of nodes within the LIS, each LIS supports a single ATM/Address Resolution Protocol (ARP) server, while all nodes (LIS clients) within the LIS are configured with the unique ATM address of the

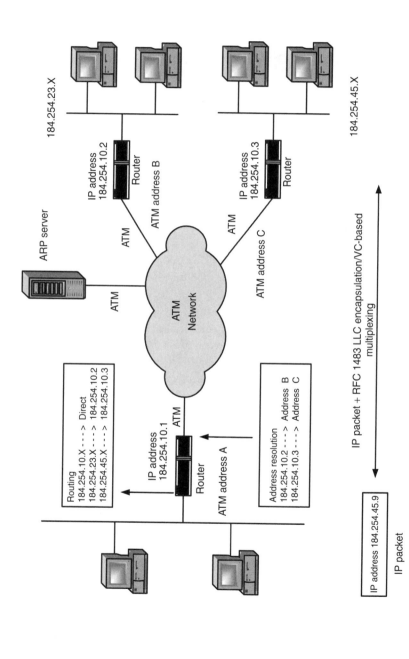

184.254.23.X

184.254.45.X

IP address
184.254.10.2

Router

ATM address B

ATM

IP address
184.254.10.3

Router

ATM

ATM address C

ARP server

ATM

ATM
Network

Routing
184.254.10.X - - - - > Direct
184.254.23.X - - - - > 184.254.10.2
184.254.45.X - - - - > 184.254.10.3

IP address
184.254.10.1

ATM

Router

ATM address A

Address resolution
184.254.10.2 - - - > Address B
184.254.10.3 - - - > Address C

IP packet + RFC 1483 LLC encapsulation/VC-based
multiplexing

IP address 184.254.45.9

IP packet

Figure 5.9 Classical IP over ATM.

ATM/ARP server. When a node source, such as a router, comes up, it first establishes a connection to the ATM/ARP server, using the server's configured ATM address. Whenever the ATM/ARP server detects a connection from a new LIS client, it transmits an inverse ARP request to the attaching client and requests that node's IP and ATM addresses, which it stores in its ATM/ARP table.

Subsequently, any node within the LIS wishing to resolve a destination IP address sends an ATM/ARP request to the server, which responds with an ATM/ARP reply if an address mapping is found. If no mapping is found, it returns an ATM negative acknowledgment response. The ATM/ARP server ages out its address table for robustness, unless clients periodically refresh their entry by responding to the server's inverse ARP queries.

Once a LIS client has obtained the ATM address that corresponds to a particular IP address, it can then set up a connection to that address using an ATM SVC. IP datagrams are sent on this ATM SVC using one of the encapsulation methods described in RFC 1483. An ATM/ARP server is required only when ATM connections are set up using ATM SVCs. ATM PVCs are preconfigured between the routers by an operator and therefore require no IP to ATM address resolution. When PVCs are used, the address resolution table in the source router simply contains the corresponding ATM VP/VC.

There are significant deficiencies in this classical IP over ATM model. For example, the protocol does not provide for IP packets destined outside the source node's IP subnet (LIS). Each LIS will have an ATM/ARP server attached to the same ATM network. Communications between two nodes on two different LISs on the same ATM network must traverse each ATM router on the path between the source and destination nodes. As a result, ATM routers become bottlenecks, and an ATM VP/VC is required between each LIS pair.

5.3.2   Next-Hop Resolution Protocol

Classical IP over ATM suffers from cut-through routes that bypass intermediate router hops when LISs send each other IP packets on the same ATM network. The continued penetration of ATM into internetworks increases the possibility that two nodes of different IP subnets connect to the same ATM network. The NHRP, defined in RFC 2332 [6], attempts to overcome this problem by building on the classical IP model but substituting a logical nonbroadcast multiaccess (NBMA) network in place of the LIS. An NMBA is a network technology, such as ATM, Frame Relay, or X.25, that permits multiple devices to be attached to the same network, but does not easily permit the use of common LAN broadcast mechanisms. NHRP operates over NBMAs.

In place of ARP servers, NHRP uses the notion of an NHRP server (NHS). Next-hop resolution clients (NHC) initiate NHRP requests of various types in order to resolve routes; for example, by mapping IP to ATM addresses. A station generally refers to a physical entity such as host or router that provides NHC or NHS functionality. Each NHS maintains next-hop resolution cache tables with IP to ATM address mappings of all nodes associated with the particular NHS, or for IP address prefixes reachable through the nodes (that is, routers) served by the NHS. NHC stations are configured with the ATM address of their NHS and then register their own ATM and IP addresses with the NHS, so that the NHS can build its cache tables.

Figure 5.10 shows a network that uses IP as the interworking protocol and ATM SVCs as the NBMA networking protocol. In this example, a source station acting as an NHC needs to route an IP packet with source IP address (S) to destination IP address (D). The source station's NHC issues an NHRP resolution request, containing the source IP address (S), the source ATM address (A), and the destination IP address (D) to NHS-1 in LIS-1. NHS-1 examines its cache and finds no entry. It examines its IP forwarding table and determines that NHS-2 in LIS-2 is the next hop on the routed path to IP address (D). NHS-1 sends an NHRP resolution packet onto NHS-2.

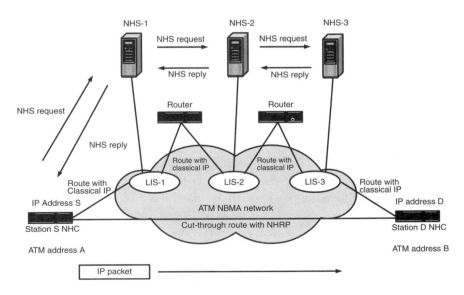

Figure 5.10 NHRP.

NHS-2 determines that there is no entry and sends an NHRP resolution request to NHS-3 in LIS-3 in the next-hop path. The station with IP address (D) is registered with NHS-3; therefore, NHS-3 returns an NHRP resolution reply containing station D's ATM address (B) to NHS-1. NHS-1 stores the address mapping between IP address (D) and ATM address (B) in its cache memory and returns the NHRP reply to station (S). Now station (S) can directly set up an ATM SVC to station (D), using the ATM address returned by the NHRP. Once the ATM SVC is set up, IP datagrams can be sent and received on this connection. Normally, the ATM SVC is torn down after a period of inactivity.

The reason the NHRP reply generally traverses the path identified in the reply is so all intermediate NHSs can also learn and cache the mapping. The next time the node requests this mapping, the NHS-1 can respond directly, without forwarding the request. However, if the NHC marks the NHRP resolution request as authoritative, then NHS-1 is forced to forward the request to the next-hop NHS without using its cached mapping.

NHRP also allows for a number of optional features, including route recording (to detect loops within the NBMA network) and fallback. Fallback is where NHSs are capable of forwarding packets along the route to a particular address. An NHS can be an intermediate forwarding point in case the actual end system is not able or willing to support direct ATM connections.

Currently, NHRP has no support for auto configuration, although this has also been a prime focus of ATM standardization efforts. It has also no support for multicast/broadcast operation. Nevertheless, NHRP will play an important role within ATM networks, particularly within the context of the MPOA, as described in Section 5.5. Interoperability between a RFC 1577/2225 [5] compliant end system and one implementing NHRP requires that the two networks be connected by a router capable of supporting both protocols.

5.3.3   IP Multicast over ATM

There is no specific IP multicast support in RFC 1577/2225 classical IP over ATM implementations. IP multicast has been hailed as the catalyst for ushering in a new generation of Internet-based services, such as VOD, broadcast video, and collaborative groupware applications. The promise of IP multicast is that it allows the simultaneous distribution of real-time voice, video, and data to a large number of subscribers, while making lower demands on available bandwidth and server resources than distribution over point-to-point IP. With IP multicast, a host can transmit the same data simultaneously to a large number of receivers that specifically request to receive that data. Like broadcast radio, IP multicast allows the receiver to tune in a specific channel and listen to information that is being sent to that channel. Unlike broadcast radio, IP multicast requires the sender to know the potential recipients.

To transmit data to a group of multicast hosts, the transmitting host sends one data stream to its nearest multicast router. This router then replicates the packet to other multicast routers within the group. The originator of the multicast data creates an IP multicast group by using one of the reserved class D multicast group/addresses, in the range 224.0.0.0 to 239.0.0.0. In Ethernet LANs, a special MAC address prefix has been reserved that allows for the mapping of class D addresses to LAN addresses.

The complex part of the IP multicast protocol is how a host indicates that it wishes to receive multicast data. A host joins an IP multicast group by sending an Internet Gateway Message Protocol (IGMP) report message, containing the IP multicast group address, to the nearest IP multicast router, as defined in RFC 1112 [7]. In this way, the multicast router keeps track of multicast group subscriptions and forwards any multicast traffic received from other multicast routers to the appropriate hosts.

IP multicast routers periodically send host membership IGMP query messages to discover which IP multicast groups have subscribers on their attached local networks. Queries are addressed to the special all-hosts group identified by IP address 224.0.0.1. Hosts respond to each received query message by sending an IGMP report message, containing the IP multicast group to which they belong. Hosts wait a random amount of time before responding with an IGMP report message so that other members of a multicast group can overhear the report and not bother sending their own. This is an important feature in IP multicasting, as multicast routers need to know only if there is at least one host on each subnetwork belonging to the group. IP multicast routers can conclude that a group has no members on this subnet when IGMP queries no longer elicit associated replies.

IP multicast over ATM described in RFC 2022 [8] defines two methods for implementing the ATM multicast feature:

1. Multicast VC mesh;
2. Multicast server.

The multicast VC mesh approach requires each node to establish its own independent PMP ATM VP/VC to its group members, forming a single multicast tree. This way, every node is able to transmit and receive from every other node, in the multicast group. This crisscrossing of ATM VP/VCs across the ATM network gives rise to the name *VC mesh*. Due to the complexity involved in configuring ATM PVCs, PMP ATM SVC capability is desired for practical deployment of a multicast ATM network.

In the multicast server approach, all nodes join a particular multicast group by establishing a point-to-point ATM VP/VC with an independent node called the multicast server (MCS). The MCS itself establishes and manages a PMP ATM VP/VC out to every node. The MCS receives IP packets from each of the nodes on point-to-point connections and then retransmits them on the PMP connection.

The relative merits of VC meshes and multicast servers depend on the trade-offs a system administrator must make between throughput, latency, congestion, and resource consumption. With a multicast server, only two ATM VP/VCs (one out, one in) are required, independent of the number of senders. The allocation of ATM VP/VC-related resources is also lower within the ATM network when using a multicast server. However, since all IP multicast traffic flows through the MCS, it becomes a single point of congestion and will be prone to bottlenecks. The advantage of a VC mesh configuration network is that it lacks the obvious single congestion point of an MCS, at the cost of requiring greater resources.

So far, we have described the two types of configurations for delivering IP multicast data traffic over ATM. Network address resolution is still required to set up ATM SVCs for the transfer of multicast IP packets. RFC 2022 [8] defines a multicast address resolution server (MARS) that is analogous to the ARP server used in the classical IP (RFC 1577/2225 [5]) over ATM model. Architecturally, the MARS is an evolution of the RFC 1577/2225 ARP server. While the ARP server keeps a table of IP/ATM address pairs for all IP endpoints in a LIS, the MARS keeps extended tables of IP to multiple ATM address mappings for all current members of the particular multicast group. It is important to know that the MARS does not take part in the actual multicasting of IP data packets; it is simply involved in ATM SVC setup.

Each MARS manages a cluster of ATM end nodes/routers. A cluster is a set of ATM end nodes/routers that choose to use the MARS to register their multicast group membership and receive updates. The MARS distributes group membership update information to cluster members over a PMP ATM VP/VC known as *ClusterControlVC*, as shown in Figure 5.11. All cluster members are leaf nodes of the ClusterControlVC.

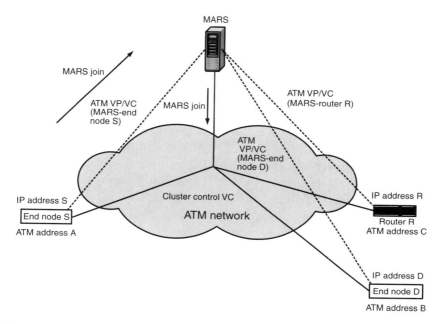

Figure 5.11 IP multicast over ATM with MARS registration.

In an MCS configuration, the MARS also needs to establish a separate *ServerControlVC* PMP ATM VP/VC out to all registered MCSs. All registered multicast servers are leaf nodes of the ServerControlVC.

By default, all MARS control messages are RFC 1483 [1] LLC/SNAP encapsulated. The unique LLC/SNAP encapsulation of MARS control messages means MARS and ARP server functionality may be implemented within a common entity and share a client-server ATM VP/VC.

End nodes at the edge of ATM networks must be configured with the ATM address of the MARS. Each end node has a dedicated ATM VP/VC connection to the MARS for address resolution and connection setup messaging, as shown in Figure 5.11. Any node wishing to join to any IP multicast group is triggered by an IGMP report message to register with the MARS server. It sends a MARS join message over its dedicated ATM VP/VC. The MARS then adds this node as a leaf to its ClusterControlVC and adds the node's address to its multicast group. The MARS emulates hardware multicast media by broadcasting the MARS join message over its ClusterControlVC to every node in its IP multicast group. All nodes receiving this message will add the new node to its PMP ATM VP/VC.

When a node wishes to send IP data packets to a particular IP multicast group and finds that it does not have any existing multicast connections, it sends a request to the MARS to obtain the ATM addresses for the IP multicast group shown in Figure 5.12. Upon receipt of these ATM addresses, the node can establish PMP ATM SVCs and transmit the data, thus emulating IP multicasting.

When a node wishes to leave an IP multicast group, it must send a MARS leave message to the MARS. The MARS removes the node's address from its IP multicast group and forwards the leave message to all nodes on its ClusterControlVC. The nodes in the IP multicast group remove the node from their individual PMP connections.

Recall that IP multicasting network multicast routers periodically send host membership IGMP query messages with destination address 224.0.0.1 to find multicast subscribers on the local network. It is much easier for the multicast routers to query the MARS instead of multicasting IGMP queries to 224.0.0.1 and incurring the associated cost of setting up an ATM VP/VC to every node in the cluster. The MARS responds with a MARS group list reply containing a list of the multicast groups within the block of IP multicast groups specified in the request. It is important to note that IP multicast group addresses are returned, not ATM addresses.

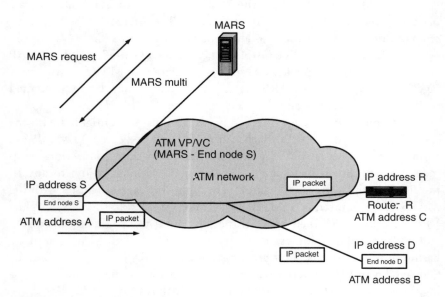

Figure 5.12 Emulating IP multicast over ATM with data.

5.4   LAN Emulation

The clip approach previously described maps IP addresses to ATM addresses using address resolution mechanisms allowing overlaying of an IP network over an ATM network. In this method, IP packets are carried across ATM networks. Another alternate method is to emulate a LAN such as IEEE 802.3 [2] Ethernet, 802.5 [9] Token Ring, or FDDI LAN over ATM. This method is known as LANE. It is important to note that LANE does not emulate the actual MAC in the associated LAN that it emulates. The LANE service has proven to be important to the acceptance of ATM, since it provides a simple and easy means for running existing LAN applications in the ATM environment.

Since LANE supports the use of multicast MAC addresses (e.g., broadcast group or functional MAC addresses), it inherently supports IP multicast better than clip. LANE was developed to enable existing applications to access an ATM network using normal IP protocol stacks. For example, IP stacks normally communicate with a MAC driver. LANE offers the same MAC driver primitives, thus keeping the IP protocol stack unchanged (i.e., making ATM transparent to the stack). This section describes the LANE protocol in an emulated local-area network (ELAN), which uses an ATM network as its backbone. Multiple ELANs may coexist simultaneously on a single ATM network.

Unlike clip, which resolves IP addresses into ATM addresses, the basic function of the LANE protocol is to resolve MAC addresses into ATM addresses. LANE actually implements a MAC bridging protocol on ATM, as shown in Figure 5.13. MAC-layer LANE is defined in a such a way that existing bridging methods like Ethernet spanning tree or Token-Ring source route can be employed as they are defined today.

The LANE protocol is generally deployed in two types of ATM-attached equipment. ATM network interface cards (NICs) that are connected to workstations or PCs implement LANE, as shown in Figure 5.13. In this case, LANE bridges ATM and LANs by interworking at the MAC layer to provide device-driver interfaces such as Open Data-Link Interface (ODI) and Network Driver Interface Specification (NDIS) to higher-level applications.

The second type of LANE device is the ATM-attached LAN switch or router. These devices, together with NIC-equipped ATM hosts, provide a virtual LAN service, where ports of the LAN switches are assigned to particular virtual LANs, independent of physical location. LANE is a particularly good fit to the first generation of LAN switches that effectively act as fast

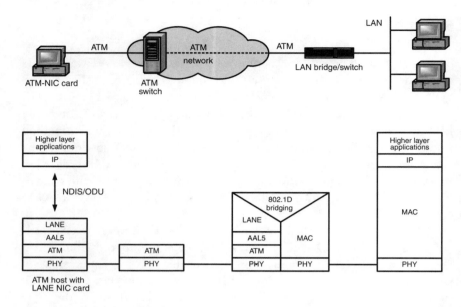

Figure 5.13 LANE protocol architecture.

multiport bridges, since LANE is essentially a protocol for bridging across ATM.

The ATM Forum's LANE 1.0 [10] specification defines operation over the ATM best effort, or UBR, service class, similar to existing LANs. The LANE 2.0 [11] specification adds QoS guarantees, giving ATM-based LANE a distinguishing characteristic over most other LAN protocols.

In order to understand the implementation of LANE, let us first look the basic logical components required in a typical configuration consisting of LAN/ATM routers/bridges and end systems. An emulated LAN is composed of a set of LAN emulation clients (LECs) and a single LAN emulation service (LE service), as shown in Figure 5.14. The LE service consists of one or more LAN emulation servers (LESs), one or more broadcast and unknown servers (BUSs), and one or more LE configuration servers (LECS).

Each LEC is part of an ATM end station such as an ATM NIC card in a workstation or PC. It represents a set of users, identified by the NIC MAC addresses. A single LEC provides one MAC service interface to one ELAN. As with any interworking implementation, there is a need to resolve the interworking protocol address to ATM addresses. The LES is responsible for resolving MAC addresses to ATM addresses. A LEC can only be connected to one LES for address resolution.

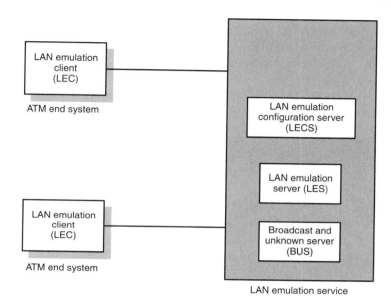

Figure 5.14 LAN emulation components.

As we have discussed in previous sections, the multicasting feature in IP networks translates to MAC multicasting in an Ethernet LAN. The BUSs handle data sent by LECs to the broadcast MAC address ("FFFFFFFFFFFF"). BUSs also handle multicast data and initial unicast data, which are sent by an LEC before the direct target ATM address has been resolved by the LES. A LEC will see only a single BUS. The BUS to which an LEC connects is identified by a unique ATM address.

The LECS provides any requesting LEC with the ATM addresses of LESs. The LECS provide the ability to assign a LEC to an emulated LAN based on either the physical location (ATM address) or the identity of a LAN destination. There is logically one LECS per administrative domain, and this serves all ELANs within that domain.

So far, we have discussed the basic components of an ELAN. The next step is to discuss how the ATM network is used as an overlay network to emulate the LAN at the MAC level. LANE defines two types of traffic that need to flow across the ATM network for a single ELAN. LECs maintain separate ATM connections for data transmission and control traffic. Communication for control traffic among the various ELAN entities (i.e., LEC, LECS, LES, and BUS) is performed over ATM VP/VCs. LANE assumes the availability of point-to-point and PMP SVCs. It is also possible, with sufficient configuration parameters, to emulate SVC functionality using PVCs and, thus, operate LANE over PVCs.

The LANE control connections are as follows:

1. *Configuration direct ATM VP/VC.* This is a bidirectional point-to-point ATM VP/VC set up by the LEC to the LECS to obtain configuration information, including the address of the LES.

2. *Control direct VP/VC.* This is a bidirectional VP/VC set up by the LEC to the LES. The LEC is required to accept control traffic on this ATM VP/VC.

3. *Control distribute ATM VP/VC.* This is a unidirectional VP/VC set up from the LES back to the LEC; this is typically a PMP connection. This is optional and is used for distributing control traffic from the LES to every LEC.

The LANE data connections are as follows:

1. *Data direct ATM VP/VC.* This is a bidirectional point-to-point ATM VP/VC set up between an LEC that wants to exchange data and another LEC. Two LECs will typically use the same data direct ATM VP/VC to carry all packets between them, rather than opening a new ATM VP/VC for each MAC address pair between them, so as to conserve connection resources and minimize setup latency. Since LANE emulates existing LANs, data direct connections will typically be UBR or ABR connections and will not offer any type of QoS guarantees. An LEC may establish additional data direct ATM VP/VCs to a destination LEC if it has traffic with particular QoS requirements and the remote LEC is able to accept such connections.

2. *Multicast send ATM VP/VC.* This is a bidirectional ATM VP/VC set up by the LEC to the BUS. It is typically a point-to-point ATM VP/VC connection. The ATM VP/VC associated with the broadcast MAC address (X"FFFFFFFFFFFF") is called the default multicast send ATM VP/VC and is used to send broadcast data to the BUS and initial data to other unicast or multicast destinations.

3. *Multicast forward ATM VP/VC.* This is a unidirectional ATM VP/VC set up to the LEC from the BUS. This is typically a PMP connection, with each LEC as a leaf. The BUS may forward data to an LEC on either a multicast send ATM VP/VC or a multicast

forward ATM VP/VC. An LEC will not receive duplicate data forwarded from the BUS (e.g., the same data will not be sent on both a multicast send ATM VP/VC and a multicast forward ATM VP/VC). The LEC must accept all data it receives on all multicast ATM VP/VCs.

During initialization, each LEC sets up a configuration direct ATM VP/VC (see step 1 in Figure 5.15) to the LECS to obtain the ATM address of the LES that is associated with the ELAN service. After obtaining the ATM address of the LES, the LEC next sets up a control direct ATM VP/VC (see step 2 in Figure 5.15) to the LES. The LEC registers its MAC address and ATM address with the LES. It may also register other MAC addresses for which it acts as a proxy, for example, other reachable MAC addresses learned by a spanning tree bridge. The LES then adds the LEC to the PMP control distribute ATM VP/VC (see step 3 in Figure 5.15). The control distribute ATM VP/VC is used to solicit an unknown ATM address (e.g., for some reason, the MAC address is behind a bridge and is not registered with the LES). If an LEC requests a MAC address, the LES can broadcast the address. The LEC that recognizes the address can respond with its ATM address.

In order to determine the ATM address of a BUS, the LEC sends the LES a request for the all-1s MAC broadcast address. The LES responds with

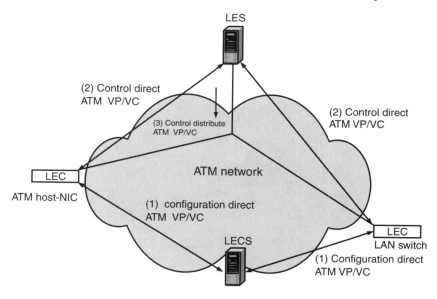

Figure 5.15 LAN emulation entities with control connections.

the BUS's ATM address. The LEC then uses this address to set up point-to-point multicast send ATM VP/VC to the BUS (see step 1 in Figure 5.16). The BUS, in turn, will add the LEC to the PMP multicast forward ATM VP/VC (see step 2 in Figure 5.16). The LEC is now ready to transfer data.

When the LEC receives a MAC frame and the ATM address is not known for the delivery of the frame, it sends a request to the LES using the control direct ATM VP/VC. While waiting for a response, the LEC may forward the frame to the BUS, which, in turn, will flood the frame to all the LECs using the multicast forward ATM VP/VC. In the meantime, when the LEC receives a response from the LES with the ATM address of the target LEC, it sets up a data direct ATM VP/VC (see step 3 in Figure 5.16). The LEC can forward all frames using this data direct ATM VP/VC.

LANE 2.0 [11] provides enhanced capabilities, including RFC 1483 [1] LLC encapsulation support for the sharing of ATM VP/VCs, as described in Section 5.2, support for ABR and other qualities of service through an expanded interface, enhanced multicast support, and support for ATM Forum multiprotocol over ATM. In order to provide QoS requirements, each LEC is permitted to establish multiple data direct ATM VP/VCs, as shown in Figure 5.17, to the same LAN destination LEC, provided the remote LEC supports this feature. When there is no QoS specified by the higher layer, the default QoS is used by the LEC for the data direct ATM VP/VC. The default is normally UBR or ABR, QoS class 0.

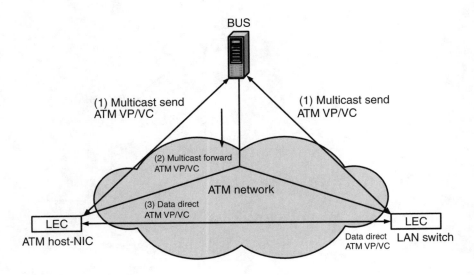

Figure 5.16 LAN emulation entities with data connections.

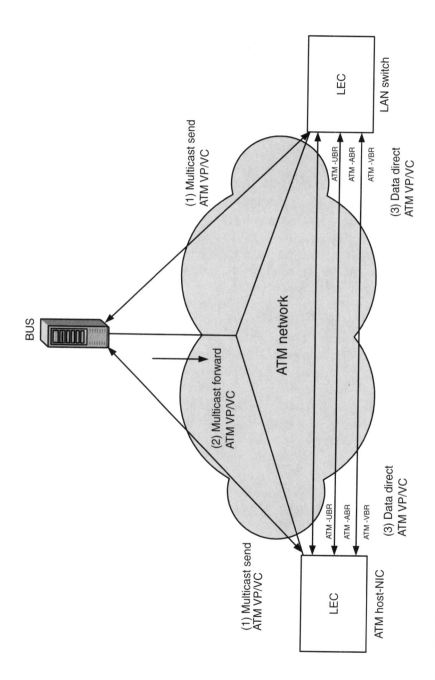

Figure 5.17 Data connections with multiple VCs with different QoS.

5.5 Multiprotocol over ATM

The ATM Forum initiated the multiprotocol over ATM (MPOA [12]) work in response to a widespread industry consensus to extend ATM to carry protocols other than IP (e.g., IPX, Apple Talk, and DECNET) and to enhance native-mode protocols. MPOA aims to efficiently transfer any protocol between different subnets in an ELAN environment. MPOA integrates the LANE and NHRP protocols and adds internetworking over dynamically established ATM SVCs without requiring routers in the data path. Thus, MPOA synthesizes bridging and routing in an ATM-based network, supporting a diverse range of link and network layer protocols.

We have discussed the NHRP, IP multicasting using MARS and LANE in previous sections. These methods allow internetwork layer protocols such as IP to operate over an ATM network by dividing it into LISs. Routers are required to interconnect these subnets, but NHRP allows intermediate routers to be bypassed on the data path, as described in Section 5.3.2. Also, while LANE provides an effective means for bridging intrasubnet data (within a LIS) across an ATM network, intersubnet (across LISs) traffic must still be forwarded through routers. Even with both LANE and NHRP, a common situation exists where communicating LAN devices are behind LANE edge devices. MPOA allows these edge devices to perform internetwork layer forwarding and establish direct communications without requiring LANE edge devices to be full-function routers. MPOA integrates LANE and NHRP to preserve the benefits of LANE, while allowing intersubnet, internetwork layer protocol communication over ATM SVCs without requiring routers in the data path. MPOA is capable of using both routing and bridging information to locate the optimal exit from the ATM cloud.

MPOA is designed with a client-server architecture. MPOA clients (MPC) and MPOA servers (MPS) are connected via LANE, as shown in Figure 5.18. The primary function of the MPC is to source and sink internetwork shortcuts. To provide this function, the MPC performs internetwork layer forwarding but does not run internetwork layer routing protocols. There are two types of MPC-based devices: MPOA edge devices contain an MPC, LEC, and a bridge port; MPOA hosts contain an MPC, LEC, and an internal host stack. These devices reside on the periphery of ATM networks, usually supporting LAN interfaces, such as Ethernet and Token Ring. MPOA edge devices are capable of bridging with other edge devices and, hence, can be part of a virtual LAN. They also have a limited amount of layer 3 processing, based on information fed from MPOA route servers.

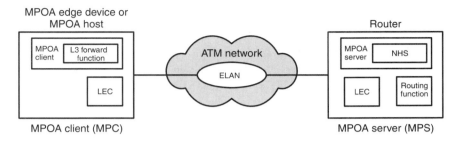

Figure 5.18 Components in an MPOA system.

An MPS is the logical component of a router that provides internetwork layer forwarding information to MPCs. An MPS includes a full NHS, with an LEC and extensions providing the layer 3 forwarding function. Its primary goal is the efficient transfer of unicast data.

All control and data flows in MPOA use ATM VP/VCs, carried over RFC 1483 [1] LLC/SNAP encapsulation. The MPOA control connections are as follows:

1. *MPC to MPS control flow.* This is an ATM VP/VC used by the ingress MPC to obtain shortcut information via MPOA resolution/reply messages.

2. *MPS to MPS control flow.* This is an ATM VP/VC for the use of MPS-MPS control using standard NHRP messaging, as described in Section 5.3.2.

3. *MPC to MPC control flow.* This an ATM VP/VC used by the egress MPC, in the event that it receives misdirected data packets, to notify the ingress MPC to invalidate its erroneous cache information.

The MPOA data connections are as follows:

1. *MPC to MPC data flow.* This is the shortcut ATM VP/VC that is set up for transferring data between the two MPCs.

2. *MPC to NHC data flow.* It is important to note that MPCs are not NHCs, as in the NHRP model described in Section 5.3.2. MPCs do not register host internetwork layer addresses with NHSs using NHRP registration. Instead, an ATM VP/VC is used to transfer unicast data between the MPC and NHC, if required.

LANE-based bridging is used by an MPC to transfer layer 3 (e.g., IP) addressed packets within a LIS. Outside the LIS, the MPS's MAC address is used as the default address for layer 3 packets. A data packet enters the MPOA system at the ingress MPC 1, as shown in Figure 5.19. By default, the packet is bridged via LANE to a router (MPS 1). MPC 1 forwards packets to MPS 1 until a threshold counter is exceeded. When the threshold is exceeded, MPC 1 is required to send an MPOA resolution request to obtain the ATM address to be used for establishing a shortcut to MPC 2. MPS 1 translates the MPOA resolution request into an NHRP resolution request, which it then forwards along the routed path. MPS 2 is the last router in the default path to the destination. It translates the NHRP resolution request to an MPOA cache imposition request and sends it to MPC 2. MPC 2 responds to the MPOA cache imposition request with an MPOA cache imposition reply. MPS 2 translates the MPOA cache imposition reply into an NHRP resolution reply, which is forwarded to MPS 1. MPS 1 translates the NHRP resolution reply to an MPOA resolution reply and returns it to MPC 1. MPC 1 establishes an ATM SVC to MPC 2. All subsequent data is then transferred using this shortcut path.

Differences in network layer protocols (IP, IPX) require MPOA components to have internetwork layer protocol-specific knowledge to perform flow detection, address resolution, and shortcut data transformations. Flow detection and address resolution require at least a minimal knowledge of internetwork layer addresses. Many internetwork layer protocols even have multiple encapsulations on a single LAN. Ingress MPC and egress MPC shortcut transformations are defined on a protocol-specific basis.

For IPX packet handling, the MPOA system supports IPX router requirements. The MPS is colocated with an IPX router and process IPX packets in conformance with IPX router requirements. For example, the ingress MPS must include a hop-count extension in MPOA resolution replies with a value as determined by the IPX routing protocol in use (e.g., 16 for RIP). Four common IPX encapsulations are in use over Ethernet (Novell proprietary), Ethernet_II, LLC, and LLC/SNAP. Ethernet_II is a frame type used in Novell networks. As a result, the MPC must support various IPX encapsulations (i.e., tagged, RFC 1483 routed, and RFC 1483 NULL) when data is sent over the shortcut route. Other media types have their own associated types of encapsulation. The ingress MPC removes the LAN encapsulation and adds the shortcut encapsulation. The egress MPC removes the shortcut encapsulation and adds the LAN encapsulation.

It is clear that a specific model for integrating ATM into today's multiprotocol networks is needed in a way that permits organizations to build

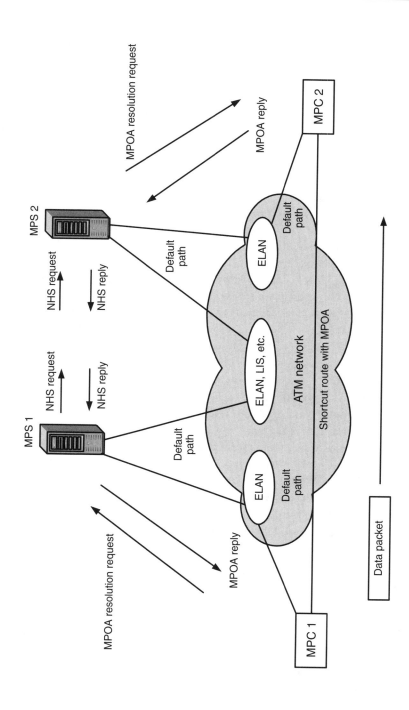

Figure 5.19 Shortcut route with MPOA.

scalable and manageable multimedia internetworks, retaining the important functionality of routers while allowing the continued use of existing Ethernet, Token Ring, TCP/IP, and IPX infrastructures. MPOA integrates LANE and HR to preserve the benefits of LANE, while allowing intersubnet network layer protocol communication over ATM SVCs without requiring conventional routers in the data path. MPOA allows the physical separation of network layer route calculation and forwarding, a technique known as virtual routing.

5.6   SS7 over ATM

SS7 over ATM enables interworking of high-bandwidth ATM networks with traditional circuit-switched SS7-based digital telecommunication networks. The convergence of voice and data switching equipment has enabled the migration of PSTN voice traffic over existing ATM data backbone networks. The interworking allows subscribers with one network to communicate transparently with subscribers in the other network over ATM.

Beginning with analog transmission and crossbar switches, telecommunication networks have evolved significantly over the past several decades. The early telecom networks used primitive signaling schemes, such as R2 (a series of specifications that refers to European analog and digital trunk signaling) and E&M (a trunking arrangement used to set up and tear down calls node by node). As transmission technology advanced, this analog transmission infrastructure was replaced by digital transmission systems. Initially, these digital transmission systems were based on PDH technology. However, with rapid developments in optical and high-speed switching technologies, SDH replaced the PDH systems. Signaling has evolved from primitive forms of CAS to common channel signaling (CCS) based on SS7. CCS has several advantages over CAS in terms of ease of implementation, centralized control, and lower equipment costs.

Circuit-switched networks restrict the efficiency of bandwidth utilization as the 64-Kbps bandwidth allocated is dedicated to the entire duration of the call. For high-speed Internet applications, circuit switching is not efficient compared with packet switching. ATM provides the features of a multiservice network to carry voice and data with guaranteed QoS. SS7 in circuit-switched networks will continue to grow, due to major investments by the telecommunications companies. In addition, a large number of areas around the world will still use voice-related services for a long time. SS7 is now widely deployed in telecom networks around the world. Interworking is required to interconnect SS7 networks with ATM networks so that transparency is provided for end customers.

Interworking is implemented in an SS7-ATM gateway at the boundary between the SS7 and ATM networks. This gateway interprets the traffic signaling going in either direction and performs the appropriate mapping and conversion between the networks. It accommodates the differences in signaling, addressing, and traffic types. The gateway can either be located at the edge of the ATM network (in an ATM switch) or as a stand-alone system located between the two networks.

In classical SS7 networks, SS7 point codes are used to address SS7 signaling messages that are routed through the network. Consider a connection between two subscribers, as shown in Figure 5.20, where subscriber A is in the SS7 network and subscriber B is in the ATM network. When subscriber A places a call to subscriber B, the call request (signaling) is routed, through a series of SS7 signaling transfer points (STPs) and signaling points (SPs), to the gateway. Circuit resources are allocated along the path through the SS7 network.

When the call request reaches the gateway, a routing table lookup at the gateway determines the (ATM) network where subscriber B lies. It also determines the signaling protocol and other parameters to be used on the ATM network, including the addressing method. The gateway performs a mapping of the received call setup request (an SS7 signaling message) to the corresponding signaling protocol message on the ATM network. To preserve service integrity, the gateway keeps as much signaling information as possible intact across the two networks. The gateway performs interworking between

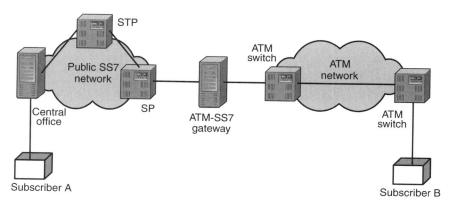

Figure 5.20 SS7-ATM interworking.

the two signaling systems and interfaces to routing and resource management. Signaling interworking is used to translate the messages and procedures from SS7 signaling protocols, such as SS7 ISDN user part (ISUP), to ATM signaling protocols, such as broadband-ISDN user part (B-ISUP). Similarly, ATM signaling protocols will be translated to SS7 signaling protocols in the other direction.

Once the signaling interworking is complete, traffic interworking is performed to map the bearer service requested by subscriber A in SS7, such as voice, 3.1-kHz audio, or $n \times 64$ unrestricted, to the appropriate ATM service type. Based on the requested SS7 bearer service, appropriate ATM QoS and traffic parameters are chosen. It is important that the gateway traffic interworking function be performed in hardware due to the speeds involved. Normally, circuit mode data ($n \times 64$ Kbps) and PCM voice use AAL1 service, and compressed voice and fax modulation/demodulation use AAL2 service.

Because the called subscriber B in the ATM network is addressed in a different way than subscriber A, address translation is performed. The translated request is then sent across to the switch into the ATM network, where it is routed to the destination switch and, ultimately, to subscriber B. User traffic can flow in each direction after the signaling is complete between subscriber A and B.

Another typical configuration connects two SS7 networks through an ATM backbone network. In this case, gateways will be implemented at both edges of the ATM network for seamless operation between the two SS7 networks.

References

[1]   Heinanen, J., "Multiprotocol Encapsulation over ATM Adaptation Layer 5," RFC 1483, July 1993.

[2]   8802-3 [ANSI/IEEE Std 802.3-1998], a bus using CSMA/CD as the access method.

[3]   IEEE 802.2 ISO/IEC 8802-2 [ANSI/IEEE Std 802.2-1998], logical link control.

[4]   ANSI/IEEE Std 802-1990 for OUI definitions, IEEE registration authority for OUI allocation.

[5]   Laubach, M., and J. Halpern, "Classical IP and ARP over ATM," RFC 1577/2225, April 1998.

[6]   Luciani, J., et al., "NBMA Next-Hop Resolution Protocol (NHRP)," RFC 2332, April 1998.

[7] Deering, S., "Host Extensions for IP Multicasting," RFC 1112, August 1989.

[8] Armitage, G., "Support for Multicast over UNI 3.0/3.1–based ATM Networks," RFC 2022, November 1996.

[9] ISO/IEC 8802-5 [ANSI/IEEE Std 802.5], using token passing ring as the access method.

[10] The ATM Forum Technical Committee, "LAN Emulation over ATM Version 1.0," af-lane-0021.000, January 1995.

[11] The ATM Forum Technical Committee, "LAN Emulation over ATM Version 2.0," af-lane-0084.000, July 1997.

[12] The ATM Forum Technical Committee, "Multiprotocol over ATM Version 1.1," af-mpoa-0114.000, May 1999.

6

Frame Relay Interworking with ATM

6.1 Introduction

The Frame Relay protocol enjoyed widespread acceptance in the network community as an access protocol and a transport protocol when the reliability of the transmission circuits improved to a level that made elaborate software-intensive error-recovery procedures in the X.25 networks too costly.

As the evolution of ATM networks continues, there is a need for Frame Relay service providers to find solutions to allow Frame Relay to interwork with ATM-based networks, thereby enabling users to have low-cost access to high-speed networks. Interworking provides connectivity between existing or new Frame Relay networks and ATM networks without changes to end user or network devices. End users are also interested in interworking Frame Relay and ATM networks to protect their capital investment in existing Frame Relay networks and eventually to migrate from Frame Relay to ATM as the offered services mature.

This chapter covers the concepts and details of the Frame Relay to ATM interworking. For Frame Relay and ATM protocols to interwork, functions such as traffic management, congestion, and PVC management for both protocols have to be mapped at the IWF. In addition, upper-layer user protocol encapsulation and address resolution need to be implemented at the IWF.

The chapter begins with the interworking scenarios, which covers the connectivity configurations of Frame Relay and ATM networks. The mapping functions at the IWF, such as bandwidth mapping, discard mapping, congestion indication mapping, and PVC status management mapping, are described in detail. The IWF is described using ATM adaptation layer AAL5 sublayer model and protocol stack. The chapter concludes with protocol encapsulation for carrying multiple upper-layer user protocols (e.g., LAN to LAN) over Frame Relay PVCs and ATM PVCs.

The Frame Relay and ATM forums have come up with standards for the interworking of the two technologies. Frame Relay/ATM network interworking FRF.5 [1] agreement is defined jointly by the Frame Relay and ATM Forums as a method to carry Frame Relay traffic over an external ATM network. This agreement allows Frame Relay traffic to be carried through another vendor's ATM/Frame Relay equipment, assuming the vendor supports the FRF.5 implementation agreement. The Frame Relay/ATM service interworking FRF.8 [2] agreement is defined jointly by the Frame Relay and ATM Forums. This agreement allows the connected devices to operate in their native ATM or Frame Relay mode. In addition, the ATM B-ISDN Intercarrier Interface (B-ICI) [3] specification was developed by the ATM Forum to address interworking with Frame Relay.

6.2 Frame Relay–ATM Interworking

The need for interworking at the IWF arises whenever the functions of one protocol have to be converted to the other for transparent end-to-end operation. The type of access protocols in operation (i.e., Frame Relay or ATM) at the two user ends, and the intervening network protocol that is assumed to be ATM in a B-ISDN network, will determine where and what type of interworking functions need to be used to allow seamless operation of user sessions.

Two different end users can be identified in a general ATM-Frame Relay interworking setup. Frame Relay customer premise equipment (FR-CPE) connects to the Frame Relay network or the IWF using the Frame Relay protocol. B-CPE or B-ISDN CPE equipment connects to the ATM network or the IWF using the ATM protocol.

Two types of interworking modes must be supported by the interworking function. In the first mode of operation, known as network interworking, the use of the B-ISDN network carrying ATM traffic is not visible to end users. There are two interworking scenarios [4] in the Frame

Relay–ATM network interworking mode of operation. Scenario 1 (see Figure 6.1) connects two Frame Relay networks/FR-CPEs using B-ISDN. The use of the B-ISDN network by two Frame Relay networks/FR-CPEs is not visible to end users. Scenario 2 (see Figure 6.1) connects a Frame Relay network/FR-CPE with an ATM B-CPE using B-ISDN such as ATM. The use of the B-ISDN network carrying ATM traffic by a Frame Relay network/FR-CPE and a B-CPE is not visible to end users. The B-CPE must have a built-in Frame Relay capability. There is no Frame Relay Q.922 [5] protocol that is implemented in the B-CPE.

The second mode of operation, known as service interworking, applies when a Frame Relay service user interworks with an ATM service user, as shown in Figure 6.2. The ATM service user performs no Frame Relay-specific functions, and the Frame Relay service user performs no ATM service-specific functions. No Frame Relay capability is needed at the ATM terminal B-CPE.

6.2.1   Mapping of Frame Relay and ATM Sessions

There are two ways of multiplexing Frame Relay over ATM. In the one-to-one mapping configuration, each Frame Relay service connection identified

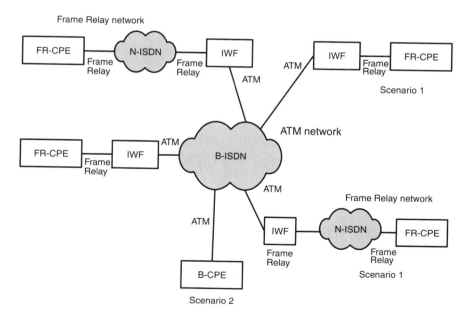

Figure 6.1 Network interworking FRF.5.

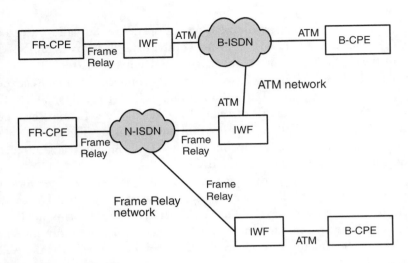

Figure 6.2 Service interworking FRF.8.

by the DLCI is mapped to a single ATM VP/VC (see Figure 6.3). Discard mapping, bandwidth mapping, congestion indication mapping, and PVC status management mapping are performed at the IWF for each Frame Relay DLCI. This method is used for both network and service interworking.

In the *N*-to-1 mapping (also known as many-to-one) configuration, a number of Frame Relay service DLCIs are multiplexed into a single ATM VP/VC (see Figure 6.4). This method is used only in network interworking. Multiplexing is an efficient way of carrying many Frame Relay DLCIs over a

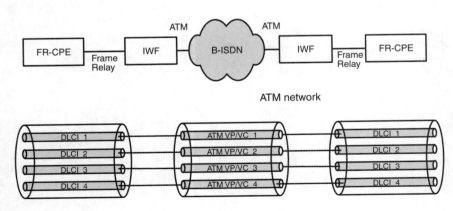

Figure 6.3 One-to-one mapping configuration.

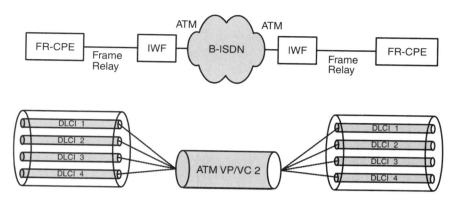

Figure 6.4 Many-to-one mapping configuration.

single ATM VP/VC if all end points are at the same destination represented by the single ATM VP/VC.

Discard mapping and congestion indication mapping are performed at the IWF for each Frame Relay service DLCI. The bandwidth mapping of the DLCIs that are multiplexed on to the single ATM VP/VC to determine the equivalent ATM bandwidth are described in Section 6.2.5. The PVC status management mapping is performed at the IWF using the reserved status signaling DLCI zero, which indicates the status of all the DLCIs for the multiplexed connection.

6.2.2   Discard-Eligible Traffic

Traffic that is nonconforming to traffic policing criteria is eligible for discard by the network. Traffic that exceeds the committed rate for the Frame Relay DLCI is marked as discard eligible. During the flow of the frames through the network(s) to its destination, the frames marked as discard eligible can be discarded for network congestion and bandwidth unavailability. The DE indication bit in the Q.922 [5] header in Frame Relay, as shown in Figure 6.5, needs to be mapped to the equivalent ATM CLP indication bit in the Frame Relay to B-ISDN direction. Similarly, in the B-ISDN to Frame Relay direction, the CLP indication is mapped to the DE bit in the Q.922 header in Frame Relay.

In the Frame Relay to B-ISDN direction, there are two mapping options that can be used. In the first option, the DE bit is mapped to the ATM CLP bit of every ATM cell generated by the Frame Relay IWF. In this option, the DE mappings are truly reflected on the ATM cells that traverse

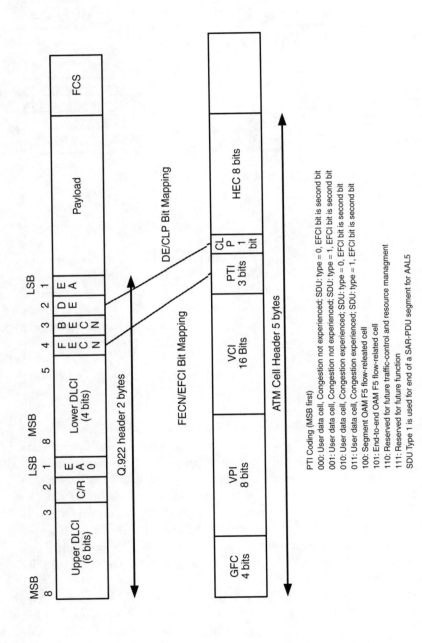

Figure 6.5 Mapping bits in protocol header.

the ATM network. In other words, DE (0) Frame Relay traffic is carried in CLP (0) ATM cells. DE (1) Frame Relay traffic is carried in CLP (1) cells.

In the second option, the ATM cells are forced to either have CLP (0) high-priority cells or CLP (1) low-priority cells, regardless of the DE bit settings. This means that the ATM traffic traverses the network as either low- or high-priority traffic toward its destination.

In the B-ISDN to the Frame Relay direction, the ATM CLP field is mapped using one of two options. The options are configurable at connection setup time. If one or more ATM cells belonging to a frame has its CLP field set to one, the Frame Relay IWF will set the DE field of the Q.922 [5] frame. The CLP bit may be set to one (tagged) along the traffic's path due to nonconformance. These tagged cells may also be discarded due to network congestion. As a result, traffic that is not discard-eligible becomes discard-eligible (DE bit set) at the receiving FR-CPE. The overall result will be the user getting a lower bandwidth than what is configured by the network operator. To minimize such problems, the network operator has the capability of configuring the CLP to DE mapping in this direction. No CLP to DE mapping is performed in the second option. If the network is prone to cell tagging, then setting the mapping to the second option can minimize the effect of unwanted discards.

6.2.3 Congestion Indication Bits

When a network node experiences traffic congestion due to excessive traffic or any resource shortage, the congestion status needs to be conveyed to the end users of the system. The end users can then take the necessary measures, such as using flow control mechanisms, to throttle the traffic and alleviate congestion. Frame Relay congestion indication in the forward and backward directions has to be mapped by the IWF to the equivalent ATM congestion indication field, as shown in Figure 6.5. Frame Relay congestion that has occurred in the same direction as the frame travels (forward direction) toward its destination is indicated by setting the FECN bit. At the same time, this Frame Relay congestion is indicated to the source end (backward direction) by the BECN bit set in frames traveling in the reverse direction to the direction of congestion. The FECN and BECN bits are set in frames by the network node when there is congestion along the route to indicate to the end users to take appropriate action. For example, a CPE receiving frames with BECN being set can then dynamically change its source transmission rate. A CPE receiving frames with FECN bit set can use congestion avoidance procedures using receiver-based flow control procedures.

Only a few end-user protocols can interpret the BECN and FECN bits. Protocols such as TCP/IP are designed for equally effective use on leased lines and in Frame Relay or ATM networks regardless of which networking protocol is used. Hence, it is not practical to incorporate functions (such as BECN and FECN) at end-user equipment that are unique to Frame Relay. However, network elements such as routers (CPE equipment) are often equipped with functions for interpreting BECN and FECN and for taking action to reduce the load through the temporary storage of data in buffers.

ATM congestion is indicated by the EFCI bit, which is part of the PTI field (see Figure 6.5) in the ATM cell header and is analogous to the FECN bit in Frame Relay. The EFCI bit is set by ATM networks when the number of cells queued in a buffer exceeds a threshold. There is no backward congestion notification in ATM, as the ATM Forum decided that the application at the destination should communicate to the application at the source to throttle using a higher-level protocol.

6.2.3.1 Congestion Indication in Forward Direction

In the Frame Relay to B-ISDN direction, the FECN in the Q.922 frame is sent unchanged in network interworking and service interworking. The EFCI of all ATM cells is always mapped to congestion not experienced. In addition, the option of mapping the FECN field in the Q.922 frame to the ATM EFCI field of every cell generated by the AAL5 segmentation and reassembly (SAR) is allowed for service interworking only.

In the B-ISDN to Frame Relay direction, if the ATM EFCI field in the last ATM cell of a segmented frame received is set to "congestion experienced" then the FECN field in the Q.922 frame will be set to indicate congestion.

6.2.3.2 Congestion Indication in Backward Direction

In the Frame Relay to B-ISDN direction for network interworking, the BECN field in the Q.922 [5] frame is set to "congestion experienced" if either of the following two conditions are met:

1. BECN is set in the Q.922 frame in the Frame Relay to B-ISDN direction.

2. EFCI is set to "congestion experienced" in the last ATM cell of the last segmented frame received in the B-ISDN to Frame Relay direction.

In the Frame Relay to B-ISDN direction for service interworking, the BECN field of the Q.922 frame has no equivalent protocol in ATM.

In the B-ISDN to Frame Relay direction for network interworking, the BECN in the Q.922 frame is sent unchanged. There is no equivalent field in ATM.

In the B-ISDN to Frame Relay direction for service interworking, the BECN field of every Q.922 frame will be set to 0, as there is no equivalent field in ATM.

6.2.4   PVC Status Management Mapping

Frame Relay PVC management procedures known as the LMI are based on periodic polling to indicate the status of Frame Relay DLCIs for PVCs. Frame Relay PVC signaling uses LMI [6] to convey end-to-end status of each Frame Relay DLCI. Frame Relay frames containing a DLCI value of zero are status frames and are used to convey the status of DLCIs. The DLCI status information is mapped to the corresponding ATM side and indications are sent to the ATM network. Likewise, indications are received from the ATM network and mapped to the corresponding status messages on the Frame Relay network/FR-CPE side. Frame Relay PVC status indications by the IWF are sent via LMI to the Frame Relay network. Thus, notification of addition, deletion, availability, and unavailability of Frame Relay DLCIs to the far-end IWF is indicated by status signaling conversion at the IWF. The Frame Relay LIV part of the LMI protocol performs the function of assuring the Frame Relay link is operational. ATM OAM cells are used to convey the status of Frame Relay link status to the far-end IWF, as shown in Figure 6.6. This is achieved by the transmission of ATM OAM AIS cells at regular time intervals (typically one cell per second) on all the affected ATM VP/VCs when a Frame Relay link fails. Upon receiving an ATM OAM AIS cell, the far end will return an ATM OAM RDI cell to indicate that the failure has been detected. Chapter 2 describes the use of ATM OAM cells in detail.

PVC status management in network interworking between the peer IWFs is accomplished by exchanging DLCI status using LMI status frames identified by the reserved status signaling DLCI zero on the configured ATM VP/VC. For the multiplexing case (N-to-1 mapping), the status signaling DLCI zero will carry the status for more than one DLCI. In the one-to-one case, the status signaling DLCI reports the status of a single DLCI.

For service interworking, the IWF receives information from the Frame Relay network/FR-CPE and maps them to the corresponding ATM indications, which are sent to the B-ISDN network. Likewise, the IWF will receive

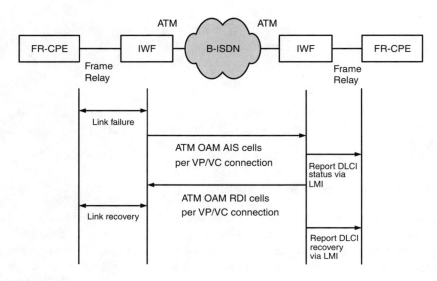

Figure 6.6 PVC status management.

indications from the B-ISDN network and map them to the corresponding Frame Relay indications. The corresponding Frame Relay status frames are then used to indicate the status to the Frame Relay network/FR-CPE. The payload for service interworking does not carry the Q.922 Frame Relay header. ATM OAM cells are used to convey the status of Frame Relay DLCIs to the far-end IWF. This is achieved by transmitting ATM OAM AIS cells at regular time intervals on the affected ATM VP/VCs when a Frame Relay DLCI becomes inactive at the IWF.

6.2.5   FR/ATM Traffic-Parameter Conversion

This section describes the Frame Relay traffic-parameter conversion to ATM traffic parameters at the IWF. Traffic bandwidth is the amount of transport resource that is needed to pass traffic between end users. It is measured in bits per second and is a precious resource that needs to be managed efficiently in the network. The mapping is done at connection establishment time for every Frame Relay DLCI that is configured on a network (see Figure 6.7). The equivalent ATM traffic parameters need to be configured on the ATM VP/VC.

The following Frame Relay traffic parameters are configurable per Frame Relay service DLCI by the network operator and used for traffic policing (ingress) and shaping (egress):

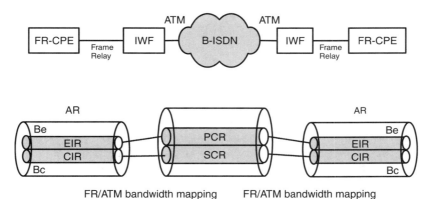

Figure 6.7 FR/ATM traffic parameter interworking.

- Bc = committed burst size (bits).
- CIR = committed information rate, Bc/ T (bps).
- Be = excess burst size (bits).

In addition to the above traffic parameters for Frame Relay, the following parameters are used to determine the equivalent ATM parameters:

- AR = access line rate (bps), which is the line rate of the Frame Relay logical port used for the DLCI. For example, if full T1 is used, then the AR will be 1,536 Kbps.
- EIR = excess information rate, Be/ T, (bps).
- CIR + EIR < = AR.

The following ATM traffic parameters are deduced from the Frame Relay traffic parameters:

- PCR = peak cell rate (cells/s).
- SCR = sustained cell rate (cells/s).
- MBS = maximum burst size (cells).

The hardware that is configured for the ATM VP/VC requires the following additional parameters:

- BT = burst tolerance (sec), derived from (MBS – 1) (1/SCR – 1/PCR).
- CDVT = cell-delay-variation tolerance (sec).

Traffic shaping lets the interworking function retain control over when to buffer or drop frames when the traffic load exceeds the guaranteed or committed values. Both ATM and Frame Relay traffic shaping are designed to transmit frames at a regulated rate, so as not to exceed some bandwidth threshold. However, Frame Relay and ATM differ in their concept of a time interval. Frame Relay PVCs transmit the Bc number of bits at any time during each time interval (T). The interval is derived from CIR and Bc. For example, assume a Frame Relay PVC with CIR of 64 Kbps and Bc of 8 Kbps:

$$Bc/CIR = Tc$$

$$8 \text{ Kbs}/64 \text{ Kbps} = 8 \ T/s$$

At each of eight Frame Relay time intervals, the Frame Relay PVC transmits 8 Kbps of data. At the end of one-second period, the PVC has transmitted 64 Kbps of data.

In contrast, ATM defines a time interval in cell units and over a sequence of received cells via the CDVT parameter, as described in Chapter 8. An ATM switch compares the actual arrival rate of adjacent cells with a theoretical arrival time. It expects a relatively consistent inter-cell gap and inter-cell arrival time. ATM switches use the CDVT value to account for arriving cell clumps with a less consistent intercell gap.

6.2.5.1 Overhead

Frame Relay frames are carried in AAL5 PDUs. As described in Section 6.3.3, there is an overhead of 8 bytes that is introduced by the AAL5 PDU trailer. The overhead calculations given by the ATM Forum's B-ICI [3] equations take the Frame Relay Q.922 [5] header and the AAL5 trailer sizes into account. ATM PVCs also introduce variable overhead of zero to 47 bytes per frame to pad the ATM AAL5 PDU to an even multiple of 48 bytes.

To deliver an equivalent CIR and EIR, the value of PCR and SCR must include the extra margin required to accommodate the overhead introduced when transferring the Frame Relay frames via an ATM network. The overhead header size for AAL5 PDU is as follows:

h1 = Frame Relay header size (bytes); whether it is 2-byte, 3-byte, or 4-byte headers, this is the overhead for the Q.922 header.

h2 = AAL type 5 PDU trailer size (8 bytes); this is the overhead for the AAL5 PDU.

h3 = Frame Relay HDLC overhead of CRC-16 and flags (4 bytes).

n = number of user information bytes in a frame.

For a Frame Relay n byte user information frame, the overhead factor (OHA) for access rate is

$$OHA\,(n)\ =\ ((n+h1+h2)/48))/n+h1+h3$$

For a Frame Relay n byte user information frame, the overhead factor (OHB) for committed and excess rate is

$$OHB\,(n)\ =\ ((n+h1+h2)/48))/n$$

6.2.5.2 Traffic Parameters for One-to-One Mapping

There are several methods used to characterize Frame Relay traffic in terms of ATM traffic conformance parameters for one-to-one mapping. The three generic cell rate algorithms (GCRAs) are intended to emulate the Frame Relay traffic parameters Bc, Be, CIR, and the Frame Relay AR. The two GCRAs can be used to either match the AR or EIR, as shown in Figure 6.8.

In the three GCRA characterizations, one GCRA (1) defines the CDVT in relation to the PCR of the aggregate CLP = 0 + 1 cell stream. The second GCRA (2) defines the sum of burst tolerance and the CDVT in relation to the SCR of the CLP = 0 cell stream. The third GCRA (3) defines the sum of burst tolerance and the CDVT in relation to the SCR of the CLP = 1 cell stream. A CLP = 0 cell that is conforming to both GCRAs (1) and (2) is said to be conforming to the configured ATM traffic contract. A CLP = 1 cell that is conforming to both GCRAs (1) and (3) is said to be conforming to the configured ATM traffic contract. A CLP = 0 cell that is not conforming to GCRA (2) above but is conforming to GCRA (1) and (3) above is considered to have the CLP bit changed to 1 and said to be conforming to the configured ATM traffic contract.

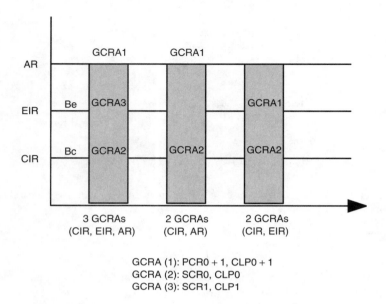

GCRA (1): PCR0 + 1, CLP0 + 1
GCRA (2): SCR0, CLP0
GCRA (3): SCR1, CLP1

Figure 6.8 Characterization of Frame Relay and ATM traffic conformance parameters.

The following ATM traffic parameters are deduced for the three GCRA characterizations in which the PCR is chosen to emulate the original Frame Relay access line rate:

1. $PCR0 + 1 = AR/8 \times [OHA(n)]$

2. $SCR0 = CIR/8 \times [OHB(n)]$

3. $MBS0 \mathrel{-}= [Bc / 8 \dfrac{(1)}{1 - CIR / AR} + 1] \times [OHB(n)]$

4. $SCR1 = EIR/8 \times [OHB(n)]$

5. $MBS1 \mathrel{-}= [Be / 8 \dfrac{(1)}{1 - EIR / AR} + 1] \times [OHB(n)]$

The two GCRAs restrict to a limitation in matching the ATM parameter to the Frame Relay EIR or the Frame Relay AR. One GCRA (1) defines the CDVT in relation to the PCR of the aggregate CLP0 + 1 cell stream. One GCRA (2) defines the burst tolerance and CDVT in relation to the SCR of the CLP = 0 cell stream.

A CLP = 0 cell that is conforming to both GCRAs (1) and (2) above is said to be conforming to the configured ATM traffic contract. A CLP = 1 cell that is conforming to GCRA (1) above is said to be conforming to the configured ATM traffic contract. A CLP = 0 cell that is not conforming to GCRA (2) above but is conforming to GCRA (1) above is considered to have the CLP bit changed to 1 and said to be conforming to the configured ATM traffic contract.

As an example, certain types of hardware may require PCR0 + 1 and CDVT for the configuration of GCRA (1) and SCR0 together with the sum of BT and CDVT for GCRA (2).

The following ATM traffic parameters are deduced for the two GCRA characterizations that provide AR matching (but does not provide a direct characterization of Be or EIR):

1. $PCR0 + 1 = AR/8 \times [OHA(n)]$

2. $SCR0 = CIR/8 \times [OHB(n)]$

3. $MBS0 \sim = [Bc / 8 \dfrac{(1)}{1 - CIR / AR} + 1] \times [OHB(n)]$

The allowed EIR can be derived using the difference between the AR and the CIR, both expressed in cells/s:

1. Allowed EIR = $8[PCR0 + 1 - SCR0]/OHB(n) = AR/8 \times [n/(n + h1 + h2)] - CIR/8$

2. Allowed Be $\sim$ = infinity

The following ATM traffic parameters are deduced for the two GCRA characterizations that provide EIR matching (but does not provide a characterization of AR):

1. $PCR0 + 1 = ((CIR + EIR)/8) \times [OHB(n)]$ (Note: CIR and EIR in bps.)

2. $SCR0 = (CIR/8) \times [OHB(n)]$ (Note: CIR in bps.)

3. $MBS0 \sim = [Bc / 8 \dfrac{(1)}{1 - CIR / AR} + 1] \times [OHB(n)]$

Table 6.1 shows the equivalent SCR values using the above equations for a Frame Relay DLCI of Bc = 1536000 bits, CIR = 1536000 bps for frame sizes ranging from 38 to 2,048 bytes. The equivalent SCR values range from 5,052 to 4,019 cells/s. If overhead of the Frame Relay header and the AAL5 trailer were ignored, the equivalent SCR values would be 4,000 cells/s. As we can see, the larger the frame size, the less significant the overhead becomes. For example, a worst-case frame size of 38 bytes at CIR of 1,536,000 bps requires an ATM bandwidth of 5,052 cells/s, which is an increase of 26%. On the other hand, a frame size of 256 bytes requires a bandwidth of 4,156 cells/s, an increase of only 3.9%. In summary, the FR-ATM overhead factor varies with packet size. Small packets result in higher padding, which results in increased overhead.

The value used for n can be based on a typical frame size, mean frame size, or worst-case scenario. If an exact packet size is unavailable for the user traffic, then an estimation can be used. For example, the average size of IP packets on the Internet is 250 bytes and this value can be used for the conversion.

These Frame Relay and ATM traffic policing/shaping parameters cannot be matched perfectly, but approximations using the recommended equations work well for most applications. In the example above, with a worst-case frame size of 38 bytes, if we take the two GCRA formula, the equations produced a difference of 26% between the ATM PVCs SCR and the Frame Relay PVCs CIR. It is possible to ignore the GCRA equations and simply configure the traffic parameters to be 15% to 25% higher on the ATM side.

Table 6.1
SCR with Frame Relay Header and AAL5 Overhead

n Frame Size	SCR No Overhead	SCR with Overhead	Percentage Increase
38	4,000	5,052	26.3
256	4,000	4,156	3.9
512	4,000	4,078	1.9
768	4,000	4,052	1.3
1,024	4,000	4,039	0.9
1,500	4,000	4,026	0.6
2,048	4,000	4,019	0.5

6.2.5.3 Traffic Parameters for Many-to-One Mapping

There are two methods used to characterize Frame Relay traffic in terms of ATM traffic conformance parameters for many-to-one mapping. Traffic characterization by PCR is used for Frame Relay traffic when the number of Frame Relay DLCIs multiplexed in an ATM VP/VC is large. Traffic characterization by PCR and SCR is used for all other Frame Relay traffic.

Traffic characterization by PCR is based on the assumption that statistical multiplexing of traffic from several DLCIs will yield a smooth traffic that can be characterized by the PCR. For example:

1. $PCR0 + 1 = $ SUM OF ALL DLCIs $((CIR + EIR)/8) \times [OHB(n)]$ (Note: CIR and EIR in bps.)

Traffic characterization by PCR and SCR does not assume the traffic smoothing effect of statistical multiplexing and accommodates the worst case:

1. $PCR0 + 1 = $ sum of all interfaces $AR/8 \times [OHA(n)]$

2. $SCR0 = $ sum of all DLCIs $CIR/8 \times [OHB(n)]$

3. $MBS0 -= [Bc\,/\,8\,\dfrac{(1)}{1 - CIR\,/\,AR} + 1] \times [OHB(n)]$

6.2.6 Examples of Network and Service Interworking

Network interworking is illustrated in the videoconferencing service example shown in Figure 6.9. The videoconferencing equipment is connected to the Frame Relay routers via Ethernet LAN. The Frame Relay routers provide access to the Frame Relay networks via DS1 physical interfaces. The IWF routers provide access to the ATM network via FRF.5 [1] interworking. The gateway for the videoconferencing equipment A is the Frame Relay router A with IP address 192.168.2.1 (subnet mask of 255.255.255.0). Similarly, the gateway for the videoconferencing equipment B is the Frame Relay router B with IP address 192.168.1.1 (subnet mask of 255.255.255.0). Frame Relay PVCs with DLCI 16 are configured between the Frame Relay routers and the IWF routers for this videoconferencing service. Similarly, ATM (PVC) VP = 0/VC = 16 is configured between IWF A and IWF B to complete the end-to-end connectivity.

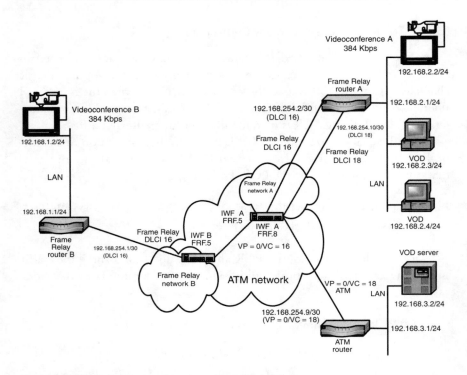

Figure 6.9 Example of network and service interworking.

In addition to the above configurations, each of the Frame Relay routers need to be configured with an IP address on the Frame Relay networking side and attached to a Frame Relay DLCI. The IP address acts as a gateway for IP traffic within the Frame Relay network. Frame Relay router A is configured with IP address 192.168.254.2 (subnet mask of 255.255.255.252) and attached to DLCI 16 (Frame Relay network A). Similarly, Frame Relay router B is configured with IP address of 192.168.254.1 (subnet mask of 255.255.255.252) and attached to DLCI 16 (Frame Relay network B). The DLCIs on Frame Relay routers A and B need not be configured with the same value (16).

IP packets destined for videoconferencing equipment B (192.168.1.2) will be sent on DLCI 16 by the Frame Relay router A to the gateway address 192.168.254.1 (Frame Relay router B). These IP packets will encounter IWF A on its way toward Frame Relay router B. At IWF A, the IP packets will be segmented into ATM cells and sent on ATM (PVC) VP = 0/VC = 16 to IWF B in accordance with FRF.5 network interworking. The IP addressing at the IWF A will not be altered. IWF B will receive the ATM cells on VP = 0/VC = 16, reassemble the IP packets in accordance with FRF.5 network interworking, and forward the packets to Frame Relay router B using Frame Relay DLCI 16. Frame Relay router B will finally forward the IP packets to the videoconferencing equipment B.

Similarly, IP packets destined for videoconferencing equipment A (192.168.2.2) will be sent on DLCI 16 by Frame Relay router B. IWF B and IWF A will use FRF.5 network interworking to transport the IP packets across the ATM network using ATM (PVC) VP = 0/VC = 16. Finally, Frame Relay router A will forward the IP packets to the videoconferencing equipment A. The Frame Relay and ATM networks are transparent to the videoconferencing equipment A and B.

The bandwidth required for videoconferencing using IP is approximately 384 Kbps. The Frame Relay DLCIs 16 (PVCs) at the Frame Relay routers A and B are configured such that the CIR is 384 Kbps, Bc is 384 Kbps, and Be is 384 Kbps. This will guarantee a bandwidth of 384 Kbps for the videoconferencing service. At the IWFs A and B, if the two GCRA characterization formula that provides EIR matching (as described in Section 6.2.5) is used, the SCR, PCR, and MBS will be 1,266 cells/s, 2,533 cells/s and 2,532 cells, respectively. The configuration for a typical FRF.5 service at the IWFs is as follows:

- AR of DS1 port at IWF A and B – 1,536 Kbps

- Frame Relay DLCI PVC – 16

- CIR – 384 Kbps

- Bc – 384 Kbs

- Be – 384 Kbs

- ATM VP for the PVC – 0

- ATM VC for the PVC – 16

Gateway configurations at Frame Relay router A such that any IP traffic bound for subnet 192.168.1.0 is routed to gateway 192.168.254.1 (DLCI 16) Frame Relay router B is as follows:

- 192.168.1.0 255.255.255.0 (subnet mask) 192.168.254.1 (Frame Relay router B as gateway).

Gateway configurations at Frame Relay router B such that any IP traffic bound for subnet 192.168.2.0 is routed to gateway 192.168.254.2 (DLCI 16) Frame Relay router A is as follows:

- 192.168.2.0 255.255.255.0 (subnet mask) 192.168.254.2 (Frame Relay router A as gateway).

Service interworking is illustrated in the VOD service configuration shown in Figure 6.9. The VOD terminal with IP address 192.168.2.3 is connected on an Ethernet LAN to a Frame Relay router A with IP address 192.168.2.1 (subnet mask of 255.255.255.0). The Frame Relay router A is connected to a Frame Relay network via a DS1 physical interface. The VOD server with IP address 192.168.3.2 is connected on an Ethernet LAN to an ATM router with IP address 192.168.3.1 (subnet mask of 255.255.255.0). The ATM router is connected to an ATM network using an OC3c physical interface.

The IWF router A at the edge of the Frame Relay network provides access to the ATM network via FRF.8 [2] service interworking. Frame Relay PVC with DLCI 18 is configured between Frame Relay router A and the IWF A for this VOD service. Similarly, an ATM (PVC) VP = 0/VC = 18 is configured between the ATM router and the IWF A router to complete the end-to-end connectivity.

The Frame Relay router A is configured with an IP address on the networking side and attached to the configured DLCI. This IP address acts as a gateway for IP traffic within the Frame Relay network. In Figure 6.9, Frame Relay router A is configured with IP address 192.168.254.10 (subnet mask of 255.255.255.252) and attached to DLCI 18 (Frame Relay network side). Similarly, the ATM router is configured with an IP address of 192.168.254.9 (subnet mask of 255.255.255.252) and attached to VP = 0/VC = 18 (ATM network side).

The gateway for the VOD terminal is Frame Relay router A (192.168.2.1). Frame Relay router A is configured so that any IP packet destined for the VOD server (192.168.3.2) will be sent on DLCI 18 with gateway address as the ATM router with IP address 192.168.254.9. At the IWF A, Frame Relay will be converted to ATM in accordance with FRF.8 and sent on VP = 0/VC = 18. The IP addressing at the IWF A will not be altered. The ATM router will receive the ATM cells on VP = 0/VC = 18 and forward the assembled IP packet onto the VOD server. Similarly, the VOD server sends IP packets destined for the VOD terminal to the ATM router (gateway IP address 192.168.3.1). The ATM router is configured so that any IP packet destined for the VOD terminal will be sent on VP = 0/VC = 18 gateway address as Frame Relay router A (IP address 192.168.254.10). At IWF A, ATM cells will be assembled and converted to Frame Relay in accordance with FRF.8 and sent on DLCI 18. The IP addressing will be unaltered. Frame Relay router A will receive the IP packet on DLCI 18 and forward it onto the VOD terminal. The Frame Relay and ATM networks are transparent to the VOD server and terminal. The bandwidth required for the VOD server and terminal using IP is typically 768 Kbps. DLCI 18 at Frame Relay router A is configured such that the CIR is 768 Kbps, Bc is 768 Kbps, and Be is 768 Kbps.

At IWF A with the two GCRA characterization that provides EIR matching formula (as described in Section 6.2.5), the SCR, PCR, and MBS will be 2,526 cells/s, PCR 5,052 cells/s, and 5,052 cells, respectively. The configuration for a typical FRF.8 service at the IWF is as follows:

- AR of DS1 port at IWF – 1,536 Kbps

- Frame DLCI – 18

- CIR – 768 Kbps

- Bc – 768 Kbs

- Be – 768 Kbs

- ATM VP – 0

- ATM VC – 18

Gateway configuration at Frame Relay router A such that any IP traffic bound for subnet 192.168.3.0 is routed to gateway 192.168.254.9 (VP = 0, VC = 18) ATM router is as follows:

- 192.168.3.0 255.255.255.0 (subnet mask) 192.168.254.9 (ATM router as gateway).

Gateway configurations at the ATM router such that any IP traffic bound for subnet 192.168.2.0 is routed to gateway 192.168.254.10 (DLCI 18) Frame Relay router A is

- 192.168.2.0 255.255.255.0 (subnet mask) 192.168.254.10 (Frame Relay router A as gateway)

In the network interworking example, IP traffic is carried across three WANs. The IWF at the edge of each WAN converts the Frame Relay to ATM and back, which adds delays (affects QoS) and unnecessary protocol conversions. The traffic parameters at each IWF need to be carefully configured so as not to affect the throughput of the service. The configuration is complex, as the operator at the Network Management Center needs to be familiar with LAN/IP, Frame Relay, and ATM networks. In the service interworking example, IP traffic is carried over two WANs using Frame Relay and ATM. Protocol conversion is done at one IWF.

6.3   Interworking Protocol Stack Organization

The basic function of the interworking protocol stack, known as the ATM adaptation layer stack, is to transport the Frame Relay payload using the ATM layer to its final destination. The payload needs to be encapsulated as AAL5 PDUs, converted (segmented) into ATM cells and sent out into the ATM network. At the destination node, the stack reassembles the received ATM cells and delivers the payload to the user (higher layer).

The AAL5 [7] provides support for higher-layer packet-based services such as Frame Relay. This section covers the AAL5 stack, which consists of the convergence sublayer (CS) and the SAR layers. The CS is subdivided into the common part CS (CPCS), and the service-specific CS (SSCS). For Frame Relay FRF.5 [1] and FRF.8 [2] interworking, the stack consists of Frame Relay service-specific convergence sublayer (FR-SSCS) [8], CPCS, and SAR sublayers. The FR-SSCS is located in the upper part of the ATM adaptation layer on top of the CPCS of AAL5. The FR-SSCS is used at the B-ISDN terminal (TE) to emulate the Frame Relay service in B-ISDN. It is also used for interworking between a B-ISDN such as ATM and a Frame Relay network.

The AAL5 SAR and CPCS sublayers as defined in B-ISDN AAL Specification I.363 [7] in conjunction with FR-SSCS (I.365.1) [8] provides the required protocol stack for Frame Relay interworking with ATM.

6.3.1   AAL5 Sublayer Model

The model shown in Figure 6.10 represents the AAL protocol structure type 5. The data unit that is passed between each sublayer is the service data unit (SDU), which is defined as a unit of interface information, the identity of which is preserved from one end of a sublayer connection to the other. The

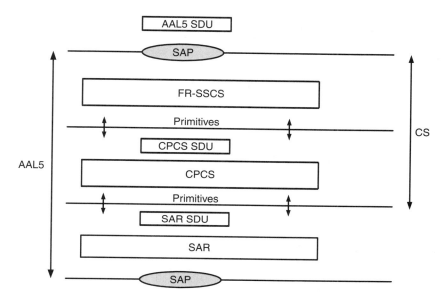

Figure 6.10 ATM AAL5 model.

AAL5 stack provides the capabilities to transfer the AAL5-SDU from one AAL5 user to another AAL5 user.

The PDU is a unit of data specified in a layer protocol and consisting of protocol control information and layer user data. A PDU from one layer becomes an SDU to the next layer.

Service access point (SAP) provides services to higher layers by passing primitives (e.g., request, indicate, response, and confirm) concerning the AAL5-PDUs. At the top, an AAL5-SAP provides services to higher layers.

SDUs are transferred between sublayers and across the AAL5-SAP. Since the CS is subdivided into SSCS and CPCS layers, the convergence specific PDU (CS-PDU) is either an SSCS-PDU or a CPCS-PDU. The AAL5 stack for Frame Relay and the ATM sublayer are shown in Figure 6.11.

Figure 6.11 shows an example where an AAL5 SDU that contains the Q.922 [5] header (the header is present only for FRF.5 [1] network interworking) and the Frame Relay payload is passed to the FR-SSCS sublayer. The CPCS layer adds padding, if necessary, and the 8-byte trailer to the SDU. The SAR sublayer segments the SDU into 48 = byte SAR PDUs. The last cell indicator (PTI) is zero for all but the last cell in the PDU. The SAR sublayer submits the SAR PDUs (48 bytes) across the ATM SAP to the ATM sublayer for transmission on the configured ATM VP/VC.

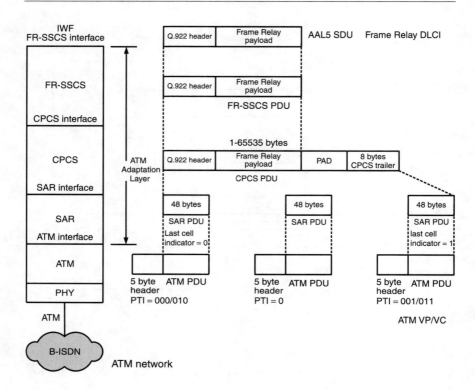

Figure 6.11 AAL5 stack for Frame Relay.

6.3.2   Frame Relay Service-Specific Convergence Sublayer

The FR-SSCS provides the bidirectional transfer of Frame Relay packets over an ATM network. It accepts a Frame Relay packet, performs the appropriate protocol conversion functions to obtain ATM cell header information from the packet, and passes the packet and header information to the lower AAL5 CPCS layer. The FR-SSCS module performs reverse protocol conversion functions to extract the Frame Relay header information from the ATM header information and, thus, outputs a reconstructed Frame Relay packet. FR-SSCS is structured so as to allow mapping a single Frame Relay DLCI to a single ATM VP/VC (one-to-one), as well as multiplexing multiple Frame Relay DLCIs (many-to-one) over a single ATM VP/VC.

The functions provided by the FR-SSCS include the following:

- *Multiplexing/Demultiplexing.* This function provides the capability to multiplex multiple DLCIs into a single ATM VP/VC.

- *Inspection of FR-SSCS PDU length.* These functions inspect the FR-SSCS PDU length to ensure that it consists of an integral number of bytes and that it is neither too long nor too short.

- *Congestion control.* These functions provide the means for the network to notify the end user, in both the forward and backward directions, that congestion avoidance procedures should be initiated. This is done by mapping the Frame Relay header's FECN and BECN bits to the EFCI bit in the ATM cell headers. In addition, the functions provide the means for the end user and/or network to indicate what frames should be discarded in preference to other frames in a congestion situation. This is done by mapping the Frame Relay header's DE bit to the CLP bit in the ATM cell header.

The peer-to-peer communication between FR-SSCS entities uses FR-SSCS-PDU. The structure is exactly the same as the Q.922 [6] Frame Relay frame when used with FRF.5 [1] network interworking without the flags, zero bit insertion, and FCS. The Q.922 Frame Relay header is removed in addition to the flags and zero bit insertion for FRF.8 [2] service interworking. The FR-SSCS will be null for FRF.8 service interworking, as it only provides the mapping of the equivalent primitives of the FR-SSCS to CPCS. Likewise, the FR-SSCS will provide only the mapping of the equivalent primitives of the FR-SSCS to the higher layer in the reverse direction.

The FR-SSCS-PDU becomes a FR-SSCS-SDU to the CPCS layer.

6.3.3 AAL5 Common Part Convergence Sublayer

One of the major functions of the AAL5 CPCS layer for Frame Relay is to convert the CPCS-SDU to a CPCS-PDU, as shown in Figure 6.11. This is achieved by appending an 8-byte CPCS-PDU trailer (see Figure 6.12) to the CPCS-SDU. A padding field provides for a 48-byte alignment of the CPCS-PDU. The CPCS-PDU trailer, together with the padding field and the CPCS-PDU payload (CPCS-SDU), comprise the CPCS-PDU. The user-to-user (UU) information is conveyed transparently between users by AAL5. The length field identifies the length of the CPCS-PDU payload so that the receiver can remove the PAD field. The CRC-32 detects errors in the CPCS-PDU.

The CPCS-PDU becomes a SAR-SDU to the AAL5 SAR layer.

The CPCS functions are performed per CPCS-PDU. The CPCS provides several functions in support of the CPCS user. The functions provided

CPCS UU (1 byte)	CPI (1 byte)	Length (2 bytes)	CRC-32 (4 bytes)

CPCS UU-CPCS: user-to-user indication, used to transparently transfer CPCS information from the originator user to the destination user.

CPI: common part indicator, used to align the CPCS-PDU trailer to the 32-bit boundary

Length: used to indicate the length of the CPCS payload (not including the PAD bytes)

CRC: cyclical redundancy check, a 32-bit error check for the entire contents of the CPCS-PDU, including the payload, the PAD field, and the first 4 bytes of the trailer

Figure 6.12 CPCS trailer in CPCS-PDU.

depend on whether the CPCS service user is operating in message or streaming mode.

1. *Message-mode service.* This service provides the transport of a single CPCS-SDU in one CPCS-PDU. When the FR-SSCS is null, the FR-SSCS-SDU is mapped to one CPCS-SDU.

2. *Streaming-mode service.* The CPCS-SDU is segmented and passed across the CPCS-interface. The transfer of these segmented CPCS-SDUs across the CPCS interface may occur separated in time. This service provides the transport of all the segmented CPCS-SDUs belonging to a single CPCS-SDU into one CPCS-PDU. An internal pipelining function in the CPCS may be applied, which provides the means by which the sending CPCS-entity initiates the transfer to the receiving CPCS-entity before it has the complete CPCS-SDU available. The streaming mode service includes an abort service by which the discarding of a CPCS-SDU partially transferred across the interface can be requested.

The CPCS user for Frame Relay, which is FR-SSCS, uses the message mode service without the corrupted data delivery option.

6.3.4   AAL5 Segmentation and Reassembly

The AAL5 SAR sublayer functions are performed on an SAR-PDU basis. The SAR sublayer accepts variable-length SAR-SDUs, which are integral multiples of 48 bytes from the CPCS and generates SAR-PDUs containing

48 bytes of SAR-SDU data. The CPCS-PDU trailer is always located in the last 8 bytes of the last SAR-PDU. The SAR sub-layer is responsible for breaking up the CPCS PDU into 48-byte parts that are passed to the ATM layer. When receiving, the SAR reassembles 48 bytes of payload into a CPCS PDU.

6.3.5   ATM Layer

The ATM layer, above the physical layer, provides the cell transfer capability. The ATM layer is responsible for appending or removing the cell header, multiplexing or demultiplexing based on VP/VC, and interpreting the cell PTI.

6.3.6   Physical Layer

The physical layer (PHY) is divided into two sublayers: the transmission convergence (TC) and the physical medium (PM).

- *Transmission convergence.* This sublayer is responsible for converting bit streams into cells, or cells into bit streams depending on direction, generating and verifying the HEC in the ATM cell header, performing cell delineation (serial mode), and performing cell rate decoupling, which sends idle cells when there is nothing to transmit. This allows the physical interface to retain synchronization. It also discards idle cells received.

- *Physical medium.* This sublayer is the interface to the transmission medium that clocks the bits. Some of the physical layer standards that exist to support ATM include DS1 (1.544 Mbps) a.k.a. T1; DS3 (54 Mbps) a.k.a. T3; OC3c (155.52 Mbps); E1 (2.048 Mbps); STS-1 (51.84 Mbps); STS-3 (155.52 Mbps); and STS-4 (622.08 Mbps).

6.3.7   FRF.5 Network Interworking Protocol Stack

FRF.5 [1] network interworking provides the capability to map a Frame Relay bearer service (FRBS) to an ATM bearer service using protocol encapsulation. The basics of network interworking are to encapsulate the Q.922 [5] frame received from the Frame Relay CPE (minus the frame check sequence) in an AAL5 PDU and forward it to the AAL5 SAR for transmission as ATM cells. In the reverse direction, the AAL5 SAR reassembles the

AAL5 PDU from the received ATM cells and finally transmits the Q.922 frame on the Frame Relay link. The interworking stack consists of a FR-SSCS and the ATM segmentation and assembly layers (CPCS and SAR), which are used to convert frames into ATM cells and reassemble the ATM cells back into frames.

FRF.5 network interworking can be deployed under two scenarios, as shown in Figures 6.13 and 6.14. In both scenarios, the use of the ATM network is not visible to the Frame Relay CPE. In these scenarios, the FRF.5 network interworking function does not modify the Frame Relay payload, which is passed transparently end-to-end. The Q.922 header may be

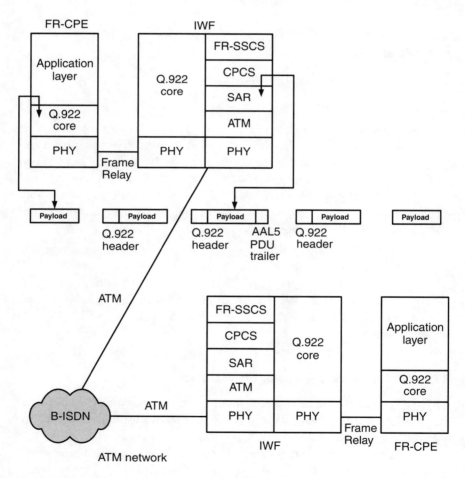

Figure 6.13 Network interworking stack FRF.5, scenario 1.

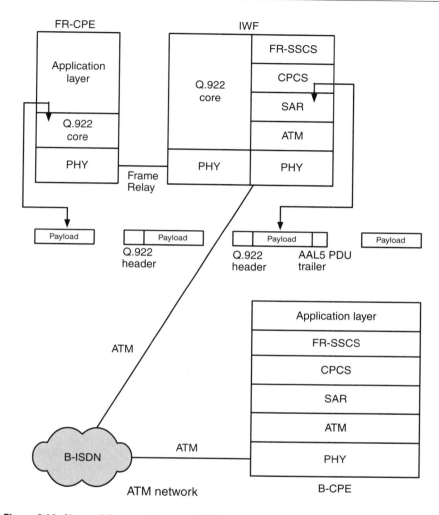

Figure 6.14 Network interworking stack FRF.5, scenario 2.

modified to relay congestion and traffic-related conditions. FR-SSCS functions are implemented at the IWF at the sending and receiving ends.

In scenario 1, the receiving end will have the Frame Relay Q.922 [5] protocol implemented to communicate with the FR-CPE (see Figure 6.13). In scenario 2, the end system is connected to B-CPE, and the Frame Relay IWF does not have the Frame Relay Q.922 protocol implemented (see Figure 6.14). The B-CPE is connected to the ATM network and must support the FR-SSCS function in its protocol stack.

6.3.8   FRF.8 Service Interworking Protocol Stack

FRF.8 [2] service interworking provides the capability to map an FRBS to an ATM bearer service using protocol mapping. The basics of service interworking are to strip the Q.922 header and frame check sequence received from the Frame Relay CPE and replace it with an ATM header, thereby mapping the Frame Relay service to an equivalent ATM service. The payload is encapsulated in an AAL5 PDU and forwarded to the AAL5 SAR for transmission as ATM cells. The B-CPE does not have any DLCI information and is transparent to the fact that the distant device is attached to a Frame Relay network.

FRF.8 service interworking can be deployed, as shown in Figure 6.15. In this scenario, the FR-CPE is communicating with a B-CPE, which does not recognize Frame Relay. The Q.922 header may be mapped to ATM congestion and traffic-related parameters on ATM cells. The FR-SSCS is replaced with a null SSCS, and the upper-layer protocols make direct use of the CPCS primitives.

6.4   FRF.8 Multiprotocol Encapsulation Mapping

Multiprotocol encapsulation in translation mode provides a flexible method for carrying multiple protocols on a single Frame Relay DLCI and ATM VP/VC. In order for the receiver to properly process the incoming AAL5 CPCS-PDU, the payload field must contain information necessary to identify the protocols. This section describes the mapping of bridged protocols (e.g., 802.3 MAC bridged), routed protocols (e.g., IP) and connection-oriented protocols (e.g., X.25, SNA) at the IWF (see Figure 6.16). As described in Chapters 4 and 5, RFC 1490 [9] and ANSI T1.617a-1993 [10] define multiprotocol encapsulation over Frame Relay and RFC 1483 [11] defines multiprotocol encapsulation over ATM, respectively.

Frame Relay uses multiprotocol encapsulation for carrying multiple protocols on a single DLCI. The Frame Relay payload is encapsulated in a RFC 1490 header. The RFC 1490 header contains a one-byte control field, one-byte NLPID field, and an optional SNAP header. The NLPID is administered by ISO and ITU. It contains values for different protocols, which includes IP, CLNP, and IEEE Subnetwork Access Protocols. For those protocols that do not have an NLPID already assigned, an NLPID value of 0x80 defines the presence of a SNAP header.

In ATM RFC 1483 encapsulation, the information to carry multiple protocols on the same ATM VP/VC is encoded in an LLC header placed in

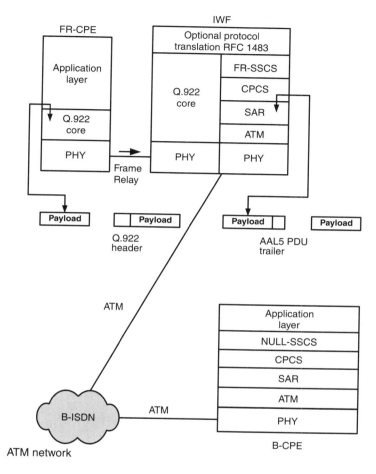

Figure 6.15 Service interworking stack FRF.8.

front of the carried PDU (see Chapter 5 on RFC 1483 LLC encapsulation) and an optional SNAP header. The LLC header consists of three one-byte fields: the DSAP, SSAP, and control. The presence of a SNAP header is indicated by the LLC header value 0xAA-AA-03.

The common optional SNAP header as defined in RFC 1490 [9, 10] (Frame Relay) and RFC 1483 (ATM) contains a three-byte OUI and a two-byte PID. The three-octet OUI identifies an organization, which administers the meaning of the following two-octet PID. For example, the OUI value of 0x0080C2 is the 802.1 organization code. The PID that follows the OUI field is administered by IEEE 802.1 [12]. The OUI and PID fields together

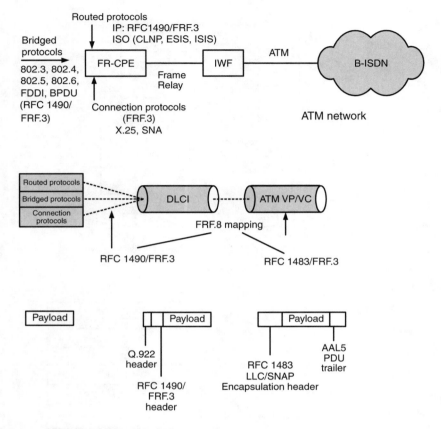

Figure 6.16 Multiprotocol encapsulation FRF.8.

identify a distinct non-ISO routed or bridged protocol. Table 6.2 shows an OUI value of 0x00-00-00, which specifies that the following PID is an Ethertype. For example, an Ethertype value of 0x08-00 is an Internet IP PDU.

As the format of the encapsulation headers for RFC 1490 [9, 10] (Frame Relay) is different from the encapsulation header defined by RFC 1483 (ATM), the FRF.8 [2] IWF needs to map between the two encapsulations, as shown in Table 6.2.

6.4.1   Bridged PDU Mapping

Bridges connect distinct LAN segments, allowing traffic to pass between segments. Bridges are used to split up large LAN segments into two smaller ones. The benefit is that there are fewer chances for collisions on the smaller

Table 6.2
Translation Mapping FRF.8

	PDU Type	RFC 1490 FRAME RELAY					RFC 1483 ATM			
		CONTROL	PAD	NLPID	SNAP HEADER OUI	SNAP HEADER PID	LLC	SNAP HEADER OUI	SNAP HEADER PID	PAD
Bridged PDU	802.3 Ethernet PDU	03	00	80 (SNAP)	0080C2 (802.1 organization code)	0001 or 0007 (NO LAN FCS)	AAAA03 (802.2 LLC header)	0080C2 (802.1 organization code)	0001 or 0007 (NO LAN FCS)	0000
	802.4 Token bus PDU	03	00	80 (SNAP)	0080C2 (802.1 organization code)	0002 or 0008 (NO LAN FCS)	AAAA03 (802.2 LLC header)	0080C2 (802.1 organization code)	0002 or 0008 (NO LAN FCS)	000000
	802.5 TOKEN RING PDU	03	00	80 (SNAP)	0080C2 (802.1 organization code)	0003 or 0009 (NO LAN FCS)	AAAA03 (802.2 LLC header)	0080C2 (802.1 organization code)	0003 or 0009 (NO LAN FCS)	0000XX
	FDDI PDU	03	00	80 (SNAP)	0080C2 (802.1 organization code)	0004 or 000a (NO LAN FCS)	AAAA03 (802.2 LLC header)	0080C2 (802.1 organization code)	0004 or 000A (NO LAN FCS)	000000
	802.6 DQDB	03	00	80 (SNAP)	0080C2 (802.1 organization code)	000B	AAAA03 (802.2 LLC header)	0080C2 (802.1 organization code)	000B	N/A
	BPDU	03	00	80 (SNAP)	0080C2 (802.1 organization code)	000E	AAAA03 (802.2 LLC header)	0080C2 (802.1 organization code)	000E	N/A
Routed PDU	Non ISO PDU	03	00	80 (SNAP)	000000	0800 IP 0806 ARP 8137 IPX	AAAA03 (802.2 LLC header)	000000	0800 IP 0806 ARP 8137 IPX	0000
	Non ISO PDU (IP)	03	N/A	CC (IP)	N/A	N/A				
	ISO PDU	03	N/A	81 (CLNP) 82 (ESIS) 83 (ISIS)	N/A	N/A	AAAA03 (802.2 LLC header)	000000	81 (CLNP) 82 (ESIS) 83 (ISIS)	N/A

XX = Don't care

segments, leaving more bandwidth for real data. The most commonly used LAN protocols, such as Ethernet/IEEE 802.3 [13], token bus/IEEE 802.4 [14], Token Ring/IEEE 802.5 [15], FDDI, and DQDB/IEEE 802.6 [16] are bridged in typical LAN configurations. These bridged protocols are encapsulated and carried over WANs such as Frame Relay using RFC 1490 and ATM using RFC 1483. The IWF supports the mapping of these bridged protocols. In addition, BPDUs and source-routed BPDUs are also supported. BPDU is the IEEE 802.1(d) [17] MAC Bridge Management protocol, which is the standard implementation of Spanning Tree Protocol (STP).

For Frame Relay RFC 1490 encapsulation, the limited numbering of the NLPID field resulted in many protocols not having specific NLPIDs assigned. Bridged packets are encapsulated using NLPID value of 0x80 indicating SNAP, and the following SNAP header identifies the format of the bridged packet. Similarly, for ATM RFC 1483 encapsulation, the presence of the SNAP header is indicated by the LLC header value 0xAA-AA-03.

6.4.2   Routed PDU Mapping

The IP is used to link the various physical networks into a single logical network. IP provides the basic packet delivery service on which TCP/IP networks are built. All protocols, in the layers above and below IP, use the IP to deliver data. TCP/IP refers to an entire suite of protocols. The suite gets its name from two of the protocols that belong to it: the TCP and the IP. TCP/IP is required for Internet connection, and its independence from physical network hardware allows it to integrate many different kinds of networks. TCP/IP can be run on Ethernet, a Token Ring, a dial-up line, an X.25 net, and virtually any other kind of physical transmission media. A common addressing scheme allows any TCP/IP device to uniquely address any other device in the entire network, even if the network is as large as the worldwide Internet. Most information about TCP/IP protocols is published as RFCs, which contain the latest versions of the specification of all standard TCP/IP protocols. The mapping of the routed protocols (see Table 6.2) that are supported at the IWF are IP, ARP and IPX PDUs.

The mapping of ISO-routed protocols that may be supported at the IWF are CLNP (ISO 8473), ESIS, and ISIS. These protocols are obsolete and are generally not supported by equipment vendors.

IP as a non-ISO protocol has an NLPID of 0xCC in the case of Frame Relay RFC 1490 encapsulation. IP is also identified as having an NLPID of 0x80 and PID of 0x08-00, as shown in Table 6.2. This format is not used; instead, IP is encapsulated like all other routed non-ISO protocols by identifying it in the SNAP header that immediately follows the LLC header in the case of ATM RFC 1483 encapsulation.

6.4.3   Connection-Oriented Protocol PDU Mapping

X.25 is an ISO standard protocol suite that specifies the interface between the DTE-user and the DCE-network in a packet-switched network. X.25

defines the lowest three layers of the ISO standard protocol stack model: the physical layer, the link layer, and the network layer.

Protocols such as SNA, which does not have a specific NLPID, use NLPID 0x08 (which indicates CCITT Q.933 [6]). The four bytes following NLPID field identify both the layer 2 and layer 3 protocols being used. Multiprotocol encapsulation (RFC 1490 [9, 10]) was enhanced to include support of the SNA protocols in FRF.3 [18, 19]. FRF.3 is used to carry SNA traffic across a Frame Relay network. FRF.3 [19] specifies how to encapsulate SNA subarea, SNA/APPN with and without HPR within the RFC 1490 multiprotocol framework. The mappings of connection-oriented protocols that are supported at the IWF are X.25/ISO 8208 packets and SNA.

6.4.3.1 Frame Relay Encapsulation of SNA Protocol

The SNA protocol does not have a specific NLPID assigned to it. The use of the Q.933 [5]/Q.2931 [20] NLPID 0x08 is used when packets of the SNA protocol are sent. The four bytes following the NLPID field include both layer 2 and layer 3 protocol identifications.

6.4.3.2 Frame Relay Encapsulation of X.25 Protocol

ANSI T1.617a-1993 [10] specifies the encapsulation of 8208/X.25 protocol using Frame Relay. The Q.922 [6] information (I) frame is for supporting layer 3 protocols, which require acknowledged data-link layer (e.g., X.25, ISO 8208). The first byte of 8208 packet also identifies the NLPID (0x01 modulo 8, 0x10 modulo 128).

6.5 FRF.8 ARP Mapping

The protocol that performs the translation of IP addresses to network addresses is the ARP. The physical networks that underlay the TCP/IP network do not understand IP addressing. The major task of the ARP protocol is to map IP addresses to the physical network's hardware addresses. ARP was initially developed for IP address resolution. The ARP for Ethernet [21], Frame Relay [9, 10], and ATM [22, 23] based networks is described in this section. The Reverse Address Resolution Protocol (RARP [24]) for Ethernet allows workstations to find dynamically their own IP addresses when they know only their hardware addresses (i.e., physical interface address). The inverse ARP (InARP) request was developed for Frame Relay [25] and ATM [22] based networks to obtain IP address of a receiving station corresponding to a newly added DLCI PVC and ATM PVC, respectively. The InARP for ATM is known as InATMARP.

Address resolution support, by transforming ARP and InARP between their Frame Relay RFC 1490 and ATM (the PVC portions of RFC 1577 [22]) formats, can only be performed when interoperating between PVCs that have been specifically configured to support FRF.8 [2] translation mode. The use of encapsulations allows ARP packets to be recognized and specially handled by the IWF.

As the Frame Relay and ATM ARP formats differ, the IWF only needs to convert between the two ARP formats. The IWF need not keep track of the end CPE IP addresses. Frame Relay and ATM stations may require resolving IP addresses dynamically. Address resolution is accomplished using the standard ARP encapsulated within a SNAP-encoded Frame Relay packet for Frame Relay or using LLC/SNAP encapsulation for ATM.

The following sections describe the various ARP implementations for Ethernet, Frame Relay and ATM networks. The ARP protocols implemented for ATM SVCs and PVCs are also described. The mapping functions between Frame Relay and ATM ARPs have to accommodate some of the differences. For example, if the IWF receives an ATM ARP with operations code 10 (NAK), the ARP packet is discarded, since Frame Relay ARP does not support that operations code.

6.5.1 ARP Ethernet

The ARP [21] function maintains a table of translations between IP addresses and Ethernet addresses. This table is built dynamically. When the ARP layer receives a request to translate an IP address, it checks for the address in its table. If the address is found, it returns the Ethernet address to the requesting node. If the address is not found in the table, ARP broadcasts a packet to every host station in the Ethernet, as shown in Figure 6.17. The packet contains the IP address for which an Ethernet address is sought. If a receiving host station identifies the IP address as its own, it responds by sending its Ethernet address back to the requesting host station. The response is then cached in the ARP table.

6.5.2 ARP in Frame Relay

Address resolution is used when a Frame Relay station needs to dynamically resolve a protocol address (IP address). The ARP [9, 10] hardware address is the Frame Relay Q.922 [6] address. The DLCI is contained within this

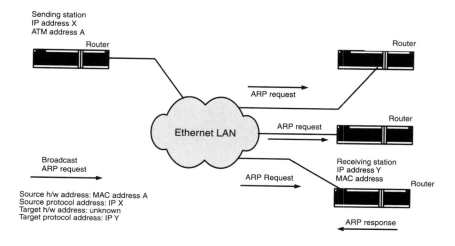

Figure 6.17 ARP—Ethernet.

Q.922 address with FECN, BECN, C/R, and DE bits set to zero. Frame Relay uses the Q.922 address contained in the Q.922 packet header for switching user data through the Frame Relay network to its final destination. For example, DLCI 50 corresponds to a Q.922 address of 0x0C-21.

DLCIs within the Frame Relay network have only local significance, and an end station will not have a specific DLCI assigned to itself. In Frame Relay networking, the DLCI is modified as it traverses the network. When the packet arrives at its destination, the DLCI has been set to the value that, from the standpoint of the receiving station, corresponds to the sending station. For example, if a station sends a packet with a Q.922 address corresponding to DLCI 70, as shown in Figure 6.18, the Q.922 address in the header will be modified from DLCI 70 to DLCI 100 by the Frame Relay network.

The ARP software in the Frame Relay station maintains a table of translations between IP addresses and Frame Relay DLCIs. When ARP receives a request to translate an IP address, it checks for the address in its local table. If the address is found, it returns a DLCI to the requesting software. If the address is not found in the table, ARP (Frame Relay) packet is sent on every DLCI that is configured for this station. Since there is no broadcast feature in Frame Relay, this is the only way it can simulate a broadcast.

Once the receiving station identifies the target protocol address (IP address) in the ARP packet as its own, it responds by sending ARP response packet by placing the receiving station's Q.922 header address (DLCI number) in the target hardware address field, as shown in Figure 6.18. When the ARP response is received at the sending station, the station will extract the Q.922 address from the Q.922 header and place it in the source hardware address field of the ARP response packet. The sending station will update its ARP table with the correct DLCI number for the required IP address.

The NLPID value 0x80 indicates the presence of a SNAP header. The OUI value of 0x00-00-000 indicates the following two bytes is a PID Ethertype. The Ethertype value of 0x08-06 indicates ARP, as shown in Table 6.2.

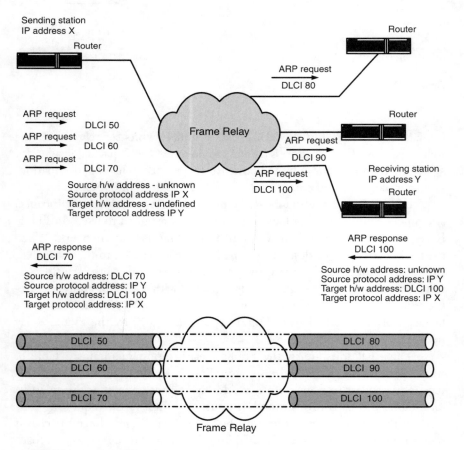

Figure 6.18 ARP—Frame Relay.

6.5.3   RARP in Ethernet

The RARP [24] translates addresses, but in the opposite sense of the function carried out by ARP. The RARP converts Ethernet addresses to IP addresses. RARP helps configure diskless systems by allowing diskless stations to learn their IP addresses. A diskless station has no disks to read its TCP/IP configuration from, including its IP address. However, every system knows its Ethernet address (MAC address) because it is encoded in the Ethernet interface hardware. The diskless station uses the Ethernet broadcast facility to ask which IP address maps to its Ethernet address. When a server on the network sees the request, it looks up the Ethernet address and if it finds a match, the server replies with the diskless station's IP address.

Bootstrap Protocol (BootP [25]) and DHCP [26] have replaced RARP in IP networks. The IP addresses are obtained from BootP and DHCP servers, respectively. DHCP servers allow LAN clients to get their IP address automatically. DHCP (RFC 1531) is based on BootP, adding the capability of automatic allocation of reusable network addresses.

6.5.4   InARP in Frame Relay

As described in previous sections, reverse ARP only allows a workstation to obtain its own protocol address (IP address). Frame Relay networks may require an IP address of a receiving station to correspond to a newly added DLCI. Clearly, RARP does not provide this feature. Due to this limitation, InARP was developed. InARP [25] allows a Frame Relay station to obtain the protocol address (IP address) of a station associated with a Frame Relay DLCI.

Unlike ARP, InARP does not broadcast requests; thus, it is more efficient. The hardware address that is the DLCI (local) of the destination station is already known, and the requesting station sends an InARP packet with source hardware address as unknown, source protocol address as IP address of the requesting station, target hardware address as the Q.922 [5] address (local DLCI) of the destination station, and the target protocol address as unknown.

The receiving station assumes that the InARP packet is destined for it. It responds by sending ARP response packet and placing the source hardware address as unknown, source protocol address as the IP address of the receiving station, target hardware address as the Q.922 local address (local DLCI), and the target protocol address as the IP address of the sending station, as shown in Figure 6.19.

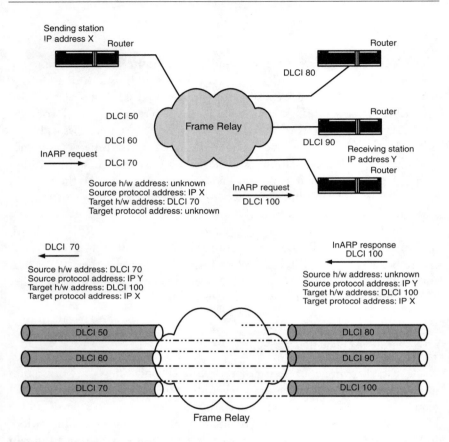

Figure 6.19 InARP—Frame Relay.

When the InARP response is received at the sending station, the station will extract the local Q.922 address (local DLCI) from the header and place it in the source hardware address field of the InARP response packet. The sending station can now update its ARP table with the correct IP address for the corresponding DLCI. Note that the source protocol address in the InARP response packet will contain the IP address of the station corresponding to the requested DLCI.

6.5.5   ARP Using ATM

Resolution of IP addresses to corresponding ATM addresses is necessary when supporting IP over ATM networks. Having to manually configure a large number of PVCs by the operator in large networks is very time consuming and not practical. An automatic configuration method using SVCs along with an ATM Address Resolution Protocol (ATMARP) server is needed. Acting as a host station to establish a connection to another host station, it must first determine the other host station's ATM address. ATMARP [22, 23] is used to resolve an IP address into an ATM address. Address resolution is performed by direct communication with a specific ARP server, rather than broadcasting ARP requests, as is done in legacy LANs. ATMARP is the same protocol as the ARP protocol, with extensions to support ARP in a unicast server ATM environment. When using SVCs, ATMARP is used for resolving IP addresses to ATM addresses. InATMARP is used for resolving ATM addresses to IP addresses. InATMARP is the same protocol as the original InARP protocol but applied to ATM networks. Use of the InATMARP protocol differs whether PVCs or SVCs are used in ATM networks. An LIS consists of a group of hosts or routers connected to an ATM network belonging to the same IP subnet. That is, they have the same IP network and subnetwork numbers. One LIS can be equated to one legacy LAN. The ATMARP server resolves ARP requests for all IP stations within the LIS. RFC 1577 [22] defines classical IP and ARP over ATM. Only the ATM ARP protocol over ATM networks is described in this section. Classical IP over ATM is described in Chapter 5.

IP stations register with the ATMARP server by placing an SVC call to the ARP server, as shown in Figure 6.20. If the LIS is operating with PVCs only, the IP station is not required to send ATMARP requests to the ATMARP server. Each host in an LIS must be configured with the ATM address of the ARP server providing ARP service for its LIS. Once the host knows its own ATM address, and the ATM address of the ATMARP server (which is manually configured in each host), it attempts to establish a connection to the ARP server, which is used to send ARP requests and receive ARP replies. When the connection to the ARP server has been established, the ARP server sends an inverse ATM ARP (InATMARP) request on the new ATM VC to learn the host's IP address to ATM address mapping in its ARP cache, as shown in Figure 6.20. The ATMARP server stores the association of a node's IP and ATM addresses in its ATMARP table. The ATM ARPserver timestamps entries and may periodically test that the IP station is still there using an InARP message after 15 minutes of inactivity. Therefore, over time, the ARP server dynamically learns the IP to ATM address mappings of all the hosts in its LIS. It can then respond to ARP requests directed toward it for hosts in its LIS.

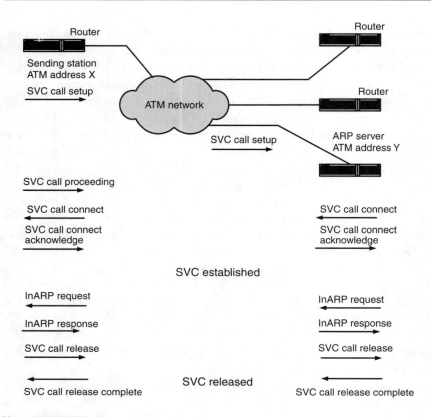

Figure 6.20 ATM ARP registration procedure.

When a host wants to communicate with another host in its LIS, it first sends an ATM ARP request to the ATMARP server containing the IP address to be resolved, as shown in Figure 6.21. When an ARP reply is received from the ATMARP server, the host creates an entry in its ARP cache for the given IP address and stores the IP to ATM address mapping. To ensure that all of the IP to ATM address mappings known by a certain host are up-to-date, hosts are required to age their ARP entries. Every 15 minutes (20 minutes by ARP server), a host must validate its ARP entries. This is done by placing an SVC call to the ATMARP server and exchanging the initial InATMARP packets. Any ARP entries not associated with open connections are immediately removed.

A host validates its SVCs by sending ARP requests to the ARP server. A host validates its PVCs, and an ARP server validates its SVCs, by sending an InATMARP request on the VP/VC. If a reply is not received, the ARP entry

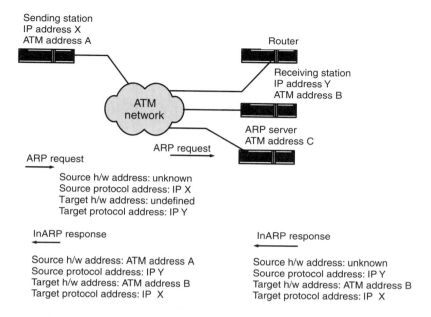

Source h/w address: unknown
Source protocol address: IP X
Target h/w address: undefined
Target protocol address: IP Y

InARP response

Source h/w address: ATM address A
Source protocol address: IP Y
Target h/w address: ATM address B
Target protocol address: IP X

InARP response

Source h/w address: unknown
Source protocol address: IP Y
Target h/w address: ATM address B
Target protocol address: IP X

Figure 6.21 ARP—ATM.

is marked as invalid. Once an ARP entry is marked as invalid, an attempt is made to revalidate it before transmitting. Transmission proceeds only when validation succeeds. If a VP/VC associated with an invalid ARP entry is closed, the entry is removed. Once the destination's ATM address is resolved, the originating station dynamically sets up an ATM SVC. After a period of inactivity, the stations take down the SVC to utilize bandwidth efficiently.

ATM addresses in Q.93B (ATM Forum UNI 3.1 [27] signaling specification) include a *calling party number information element* and a *calling party subaddress information element*. These information elements are mapped to source ATM number and source ATM subaddress, respectively. Similarly, the ATM Forum defines a *called party number information element* and a called party subaddress information element. These information elements are mapped to target ATM number and target ATM subaddress, respectively.

RFC 1577 [22] specifies that implementations must support IEEE 802.2 logical control/subnetwork attachment point (LLC/SNAP) encapsulation, as described in RFC 1483 [11]. LLC/SNAP encapsulation is the default packet format for IP datagrams. The LLC header value 0xAA-AA-03 indicates the presence of a SNAP header. The OUI value of 0x00-00-000 indicates the following two-bytes is an Ethertype. The Ethertype value of 0x08-06 indicates ARP, as shown in Table 6.2.

6.5.6   InATMARP Using ATM PVC

When using ATM PVCs, InATMARP is used for resolving ATM VP/VC to IP address resolutions. RFC 2390 [25] specifies InARP as a means for stations to automatically learn the IP address of the stations on the other end of an ATM VP/VC. All stations using PVCs are required to use the InATMARP as specified in RFC 1577 [22] on those ATM VP/VCs using LLC/SNAP encapsulation. The stations using PVCs are manually configured for what PVCs it has and, in particular, which PVCs are being used with LLC/SNAP encapsulation. The station sends an InATMARP request containing the sender's IP address over the ATM VP/VC. The station on the other end of the ATM VP/VC then responds with its IP address, establishing an association between the IP addresses of the pair and the ATM VP/VC on each ATM interface. In a strict ATM PVC environment, the receiving station will infer the ATM VP/VC from the ATM VP/VC on which the InATMARP request or response was received.

The requesting station with IP address X sends an InATMARP packet over the ATM network on VP/VC 70. This appears on VP/VC 100 on the receiving station whose IP address is Y. The receiving station responds with IP address Y using InATMARP response packet on VP/VC 100. The sending station receives this response on VP/VC 70, thus the stations now know of each other's IP addresses and can reach each other by transmitting on ATM VP/VC 70 and 100, respectively.

The continued penetration of ATM into internetworks increases the possibility that two nodes of different IP subnets connect to the same ATM network. In place of ARP servers, the NHRP uses the notion of an NHRP server (NHS). NHCs initiate NHRP requests of various types in order to resolve routes; for example, by mapping IP to ATM addresses. Chapter 5 details the concepts of NHRP that use a network technology, such as ATM, which permits multiple devices to be attached to the same network but does not easily permit the use of common LAN broadcast mechanisms.

6.6   Summary

In this chapter, the interworking agreements between Frame Relay networks and ATM (B-ISDN) networks as defined in Frame Relay Forum documents

FRF.5 [1] (network interworking) and FRF.8 [2] (service interworking) have been detailed. The importance of the IETF encapsulation methods outlined in RFC 1490 [9, 10] (IP over Frame Relay) and RFC 1483 [11] (IP over ATM) as being the central documents to the formats used in Frame Relay to ATM interworking has been observed. In addition, currently, interworking methods have been developed and implemented for systems where only PVCs are employed. More flexible interworking arrangements that include SVCs are under development, and practical implementations are not yet common. Various ARPs that would participate in a more generic interworking scenario are also described in this chapter.

References

[1] O'Leary, D., "Frame Relay/ATM PVC Network Interworking Implementation Agreement," FRF.5, December 20, 1994.

[2] O'Leary, D., "Frame Relay/ATM PVC Service Interworking Implementation Agreement," FRF.8, April 14, 1995.

[3] ATM Forum Technical Committee, "BISDN Inter-Carrier Interface (B-ICI) Specification Version 2.0 (Integrated)," af-bici-0013.003, December 1995.

[4] ITU-T Recommendation I.555 Series 1, "Frame Relaying Bearer Service Interworking," June 1993.

[5] ITU-T Recommendation Q.922, "ISDN Data Link Layer Specification for Frame Mode Bearer Services," 1992.

[6] ITU-T Recommendation Q.933 bis, "Digital Subscriber Signaling System No. 1— Annex A," October 1995.

[7] ITU-T Recommendation I.363, "B-ISDN ATM Adaptation Layer (AAL) Specification," March 1993.

[8] ITU-T Recommendation I.365.1, "Frame Relay Service-Specific Convergence Sublayer (FR-SSCS)," November 1993.

[9] Bradley, T., et al., "Multiprotocol Interconnect over Frame Relay," RFC 1490, July 1993.

[10] ANSI T1.617a, "Multiprotocol Encapsulation over Frame Relay ANNEX F," 1994.

[11] Heinanen, J., "Multiprotocol Encapsulation over ATM Adaptation Layer 5," RFC 1483, July 1993.

[12] ANSI/IEEE Std 802-1990 for OUI Definitions, IEEE Registration Authority for OUI Allocation.

[13] ISO 8802-3 [ANSI/IEEE Std 802.3-1998], a bus using CSMA/CD as the access method.

[14] ISO/IEC 8802-4 [ANSI/IEEE Std 802.4-1990], a bus using token passing as the access method.

[15] ISO/IEC 8802-5 [ANSI/IEEE Std 802.5], using token passing ring as the access method.

[16] ISO/IEC 8802-6 [ANSI/IEEE Std 802.6], using distributed queuing dual bus as the access method.

[17] ISO/IEC 15802-3 [ANSI/IEEE Std 802.1D-1998], media access control (MAC) bridges.

[18] Cherukuri, R., "Multiprotocol Encapsulation Implementation Agreements," FRF.3, June 1995.

[19] "Multiprotocol Encapsulation Implementation Agreement (MEI)," FRF.3.1 Frame Relay Forum Implementation Agreement, June 1995.

[20] ITU-T Recommendation Q.2931 B-ISDN, "Application Protocols for Access Signaling," February 1995; Amendment 1, June 1997.

[21] Plummer, D., "An Ethernet Address Resolution Protocol," RFC 826, November 1982.

[22] Laubach, M., "Classical IP and ARP over ATM," RFC 1577, January 1994.

[23] Laubach, M., and J. Halpern, "Classical IP and ARP over ATM," RFC 2225, April 1998.

[24] Finlayson, R., et al., "A Reverse Address Resolution Protocol," RFC 903, June 1984.

[25] Bradley, T., C. Brown, and A. Malis, "Inverse Address Resolution Protocol," RFC 2390, September 1998.

[26] Bradley, T., and C. Brown, "Inverse Address Resolution Protocol," RFC 1293 , January 1992.

[27] Croft, B., and J. Gilmore, "Bootstrap Protocol (BootP)," RFC 951, September 1985.

Selected Bibliography

ITU-T Recommendation I.370 ISDN User-Network Interfaces, "Congestion Management for the ISDN Frame Relaying Bearer Service," 1991.

"Frame Relay Multiprotocol Encapsulation Implementation Agreement," FRF.3.2 Frame Relay Forum Implementation Agreement, April 2000.

7

CES Service Interworking with ATM

7.1 Introduction

CES refers to a service that transports CBR traffic over an ATM network. Circuit types defined by the ATM Forum specifications [1] as supporting CES service include structured and unstructured forms of T1 and E1, and the unstructured form of T3 and E3. In an unstructured form of service, the entire bandwidth carried by the interface is transported transparently over the ATM network. It can essentially be thought of as emulating the T1 or E1 leased line service. In structured form CBR service, fractional T1 or E1 bandwidths are transported to another network point over the ATM network. Figures 7.1 and 7.2 illustrate typical configuration of networks supporting structured and unstructured CES (UCES) services.

In this chapter, we will cover the following aspects of CES services:

1. Interworking is based on standard formats to carry circuit data in ATM cells. The cell coding technique fundamental to the support of CES services is based on the generic ATM AAL1 specifications defined in ITU-T Recommendation I.363.1 [2]. The cell formats specific to CES services are defined in the ATM Forum CES interoperability specification af-vtoa-0078 [1]. Both will be described in this chapter.

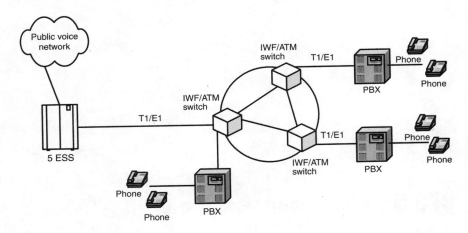

Figure 7.1 Example of the use of UCES.

Router CES traffic: router A to router B
CES fax traffic: fax A to fax B

Figure 7.2 Example of the use of SCES.

2. When circuit-switched data is converted to packet form and transported over the ATM network, synchronization of transmit clocks between the near-end user and the terminating interworking function side transmit clock becomes a key system issue. Clocking arrangements are used to ensure that circuit timing is maintained at the two ends of a circuit that spans a cell-based ATM network that does not itself use any circuit-related timing for transport. Clock timestamp and recovery methods will be covered in this chapter.

3. Functions of the AAL1 layer are typically done completely in silicon. When a CES service is provisioned, the hardware supporting the interworking of CBR data and ATM is configured according to the nature of the CES connection (e.g., structured or unstructured, no other software participates in assisting the data flow). Implementation issues associated with CES services will also be covered in this chapter.

7.2 Overview of AAL1

The basic architecture that is used to define the functional layers of the ATM stack is shown in Figure 7.3. The application that will reside above the AAL

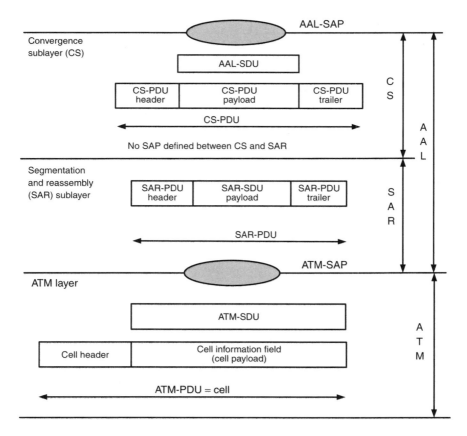

Figure 7.3 Functional layers of ATM stack.

in supporting the CES will be the stream of information at constant bit rate carried by different service subtypes encompassed by the CES service, namely unstructured T1/E1 or fractional T1/E1 services. The adaptation layer has the responsibility to convert CBR bit stream to 48-byte ATM cells for delivery to the other end of the connection. In this process, the layer uses a predefined structure (cell coding) for the ATM payload. This structure includes a facility to transfer timing information and also incorporates a mechanism to ensure the integrity of cell data and the ability of the peer layer to detect a loss of cells.

The AAL consists of two sublayers: a CS and an SAR, as described in Sections 7.2.1 and 7.2.2.

7.2.1   The Convergence Sublayer

In the transmitting direction, the CS layer will receive from the AAL user CBR information that can be characterized as consisting of single-bit AAL-SDUs. The CS layer will accumulate AAL-SDUs to construct a 47-byte SAR-PDU for delivery to the SAR layer. The 47-byte payload can have two different formats, the P format where the user information forms 46 bytes with a 1-byte pointer field making up the 47-byte SAR-PDU, and the non-P format where the 47-byte SAR-PDU is built exclusively from AAL user information.

The CS also maintains a modulo 8 sequence count of the SAR-PDU and provides this count to the SAR layer for incorporation into the SAR-PDU header. The count is used by the CS layer to identify missing cells or identify misinserted cells.

The reassembly unit (referred to as the AAL receiver), in addition to detecting lost or misinserted cells during sequence number processing, can also employ buffer-fill level to detect lost or misinserted cells. These mechanisms are necessary to maintain bit count integrity where the layer has to maintain the counts of incoming bit stream (into segmenting) and outgoing bit stream (out of reassembly) to be the same whenever possible. If buffer underflows, or no cell arrives in a predetermined interval, dummy bits are filled or inserted. Similarly, if buffer overflows or a nonconforming cell arrives, the bits or octets have to be discarded. The value of bits to be inserted depends on the service being supported by the AAL1 layer. When the CS layer is supporting circuit transport at 1.544 Mbps and 2.048 Mbps, as per Recommendation G.702 [3], the inserted dummy bits shall all be 1s. Also, when the AAL1 supports voice band signal transport, there is no need to detect misinserted cells. However, the receiving AAL entity must detect and compensate for lost cells to maintain bit count integrity and minimize delay, thereby alleviating echo performance problems.

7.2.2 SAR Sublayer

The SAR sublayer performs the mapping between the PDUs of the CS and SAR sublayers, conveys the existence of a CS function to the peer CS entity, provides sequence numbering support to enable the CS to detect lost or misinserted PDUs, and provides error protection to the sequence count value and other fields.

The main function of the layer in the transmitting direction is to create a 4-bit sequence number (SN) field and a 4-bit sequence number protection (SNP) field. SAR includes this information in a header byte, which it adds to the 47-byte payload delivered from the CS layer.

A SAR-header byte is constructed from the sequence count (3-bits) and a single CS-indication (CSI) bit provided by the CS layer, as shown in Figures 7.4(a–c).

A 3-bit CRC on the first 4 bits of the header byte is computed using generator polynomial $x \times 3 + x + 1$, and the resulting 7-bit code word is used to create another even parity bit.

7.2.3 Cell Coding: Structured Data Transfer Method

A structured form of cell coding is used to handle octet-based data. The CS uses two different formats, P format and non-P format, as shown in Figure 7.5. Non-P format is used in all SAR-PDUs that have an odd

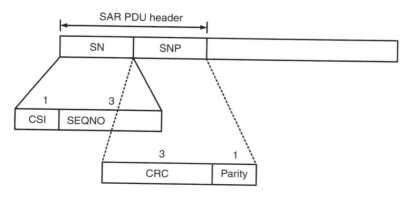

Figure 7.4(a) Format of AAL1 SAR PDU.

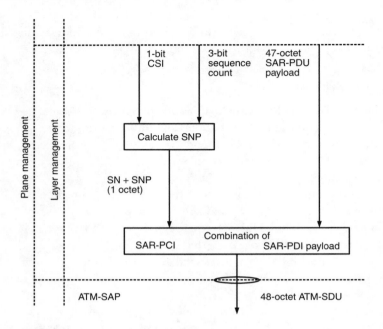

Figure 7.4(b)   Functional model of SAR at transmit side.

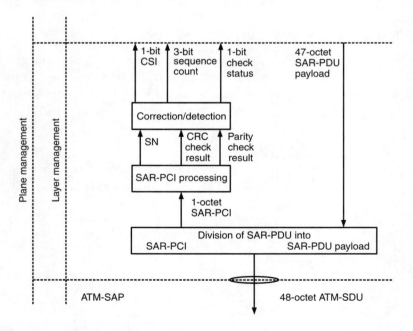

Figure 7.4(c)   Functional model of SAR at receiving side.

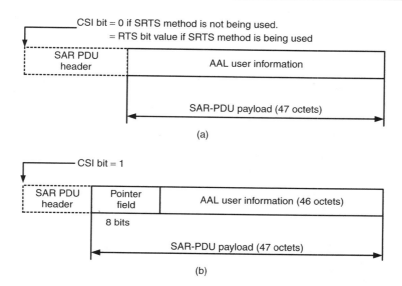

CSI bit = 0 if SRTS method is not being used.
= RTS bit value if SRTS method is being used

Figure 7.5 Non-P and P format of SAR-PDU payload.

sequence number (1, 3, 5, or 7. Even-numbered cells can be either P or non-P format. In the non-P format, all 47 octets carry user data.

In the P format, the first octet of the SAR-PDU is always a pointer value, followed by 46 bytes of user data. The pointer byte essentially defines the offset in number of bytes of the beginning of the next logical data block from the pointer byte. As we have noted earlier, since only the even-numbered cells can have the P format, the offset value needs to also be able to denote the beginning of a new data block at any point in the following non-P format cell payload. The valid range of the pointer byte is therefore 0–93, allowing for the offset that can cover the P format cell (46 bytes) added to the payload length in the adjacent non-P format cell (47 bytes).

In addition, the P format is used only once in the 8-cell sequence numbered 0 to 7. If the block is large enough that no end-of-block edge falls within a particular 8-cell cycle, a dummy pointer (with value 127) is used in the last possible opportunity in a cycle, which is cell with sequence number 6.

7.3 Clock-Recovery Methods

In CES services, clock synchronization between the near-end and far-end equipment will be necessary to successfully set up a connection through an ATM network. Two different methods of clock synchronization are available:

1. *Synchronous mode:* In this mode, timing to the user equipment is provided by an external source such as a GPS clock or from the access equipment. The timing source is generally derived from a primary reference source (PRS), as illustrated in Figure 7.6(a). In synchronous mode, one can assume that equipment clocks at both ends are locked to the clock from a PRS.

2. *Asynchronous mode:* In asynchronous mode, service clocks supporting CES services are not locked to a primary source. This forces the destination IWF's transmit clock to continuously adjust to maintain synchronism with the clock of the sending end. Two different asynchronous clocking methods are available, the adaptive clock method and the synchronous residual timestamp (SRTS) method.

The SRTS method is more accurate as it meets the jitter-and-wander requirements defined in G.823 [4] and G.824 [5], and it is generally used in public networks. The adaptive clock method is less accurate and, hence, is generally used for those services that need to comply with jitter requirements but do not need to comply with wander requirements.

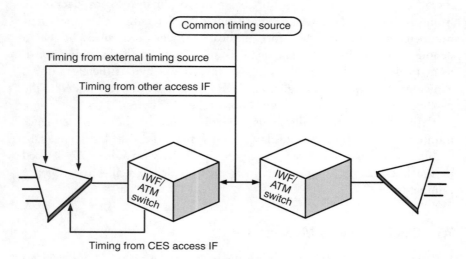

Figure 7.6(a) Synchronous mode clocking.

Both methods are shown in Figures 7.6(b, c) and discussed in the following sections.

Forum specifications recommend that structured CBR services use synchronous clocking and unstructured CBR services use either synchronous or asynchronous clocking.

7.3.1 Adaptive Clock Recovery

In this form of asynchronous mode clock recovery, there is no common clock between the sending and receiving ends of the CES link and no timing

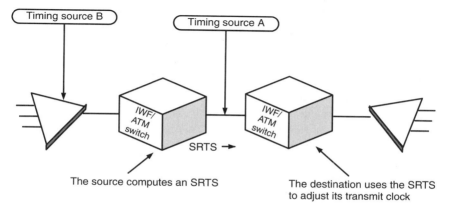

Figure 7.6(b) Asynchronous mode clocking (SRTS).

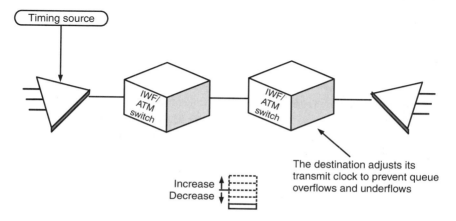

Figure 7.6(c) Asynchronous mode clocking (adaptive).

information is transmitted from the source end. The adaptive method is implemented at the receiving adaptation layer to derive its transmit clock by measuring the queue fill levels of the buffer that holds the data received from the source.

Adaptive recovery implementation techniques are not standardized, but all are based on the premise that the amount of transmitted data gives a measure of the source frequency. By averaging the data received over a larger period of time compared to the cell delay variation (CDV) value, the adaptive recovery scheme reduces the impact of CDV on the recovered clock frequency.

One commonly used method for adaptive recovery employs current-fill levels in the reassembly (receive buffer) to control the frequency of the service clock. The method first designates a median level for buffer fill in the reassembly buffer at the destination end. A level above the median (high mark) and another level below the median (low mark) are chosen [see Figure 7.6(b)], and the clock frequency is adjusted to keep the current fill level controlled to remain within the high and low marks. The current fill level will be used as a control input to the phase-locked loop that determines the frequency of the service clock. If the buffer depth exceeds the higher mark, the clock transmit frequency will be increased to drain additional cells. If the depth falls below the low mark, then the clock transmit frequency will be reduced to allow the fill level to rise above the low mark.

The adaptive method can potentially introduce wander if the CDV contains a low frequency component. This method is generally used only in private networks where clocking requirements are not very stringent, as private network owners can opt for a lower cost solution with increased network transmission error rates.

7.3.2 SRTS

The SRTS method of asynchronous mode clock recovery involves the use of a network-wide synchronization signal, traceable to a PRS. The difference between this synchronization signal and the local service clock frequency of the source IWF is measured and sent as a timestamp value in the adaptation layer header to the reassembly IWF at the destination end. The service clock for the destination IWF is then reconstituted using the timestamp and the same PRS synchronization signal available at the destination.

The SRTS technique can be illustrated with the aid of Figure 7.7(a, b).

A constant time period (T) that covers the 8-cell sequence serves as the reference time frame for the SRTS-based synchronization. The 8-cell

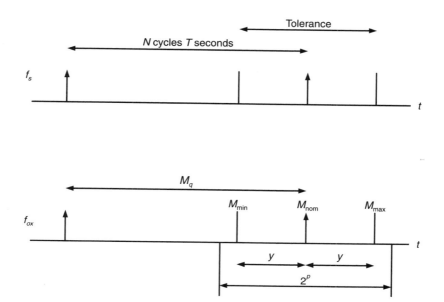

Figure 7.7(a) Concept of synchronous residual timestamp.

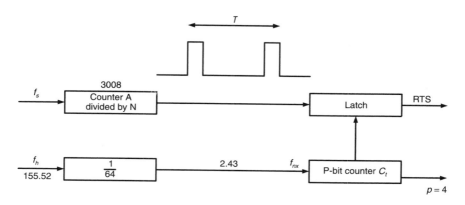

Figure 7.7(b) Generation of residual timestamp.

sequence will represent an incoming bit sequence of $N = 8$ cells/period $\times$ 47 bytes/cell $\times$ 8 bits/byte $= 3,008$ bits/period (non-P format sequence) at the service clock frequency f_s. A derived reference clock frequency f_{ox} is generated to be greater than f_s and less than $2f_s$ by dividing the network clock frequency of 155.52 MHz by 2^k. With the above parameters, one can calculate the number of f_{ox} clock cycles that will fall within the time period T, $M_{nominal} = N \times f_{ox}/f_s$. Then, if the expected variation of the f_s within the period T is $+/- e\%$, then in terms of the number of f_{ox} cycles this variation will represent is given by $y = N \times f_{ox}/f_s \times e$.

If y is rounded up to the next integer, one needs a size of p bits to represent this difference, given by $2^{(p-1)} > y$.

To represent a tolerance of 200×10^{-6}, the above equations yield a value for p as 4 bits. Figure 7.7(b) shows a hardware scheme that generates the residual time stamp (RTS) where the counter is clocked by f_{ox}, and the residue is latched every period T.

The 4-bit SRS is carried in the CSI bit of the SAR-header, one bit in each odd-sequence (1, 3, 5, and 7) cell. The receiver with the knowledge of the $M_{nominal}$ (fixed for a given N) and the 4-bit RTS can construct a reference signal that will be used to drive the phase-locked loop that generates the destination IWF service clock.

Also, we need to note that when data carried in the CES session is P-format in the structured data transfer (SDT) method, the N is reduced to 3,000 to account for the one-bit overhead that is used exactly once per 8-cell cycle.

7.4   Interworking Scheme for Unstructured CES

For applications that use the entire bandwidth, including the framing bits of DS1 (1.544 Mbps), E1 (2.048 Mbps), and J2 (6.312 Mbps), UCES provides a convenient way to transport data over an ATM network. Currently, this scheme is used in a large number of applications, including public voice networks and PBX-based voice systems.

7.4.1   UCES Protocol Stack

Figure 7.8 illustrates the protocol stack organization of the interworking function supporting a UCES service. The shaded section represents the physical interface side of the circuit being emulated and serves as the user entity for the ATM network side stack supporting the AAL1 adaptation layer.

With unstructured service, this interworking function simply provides the mechanism to transfer bits between the physical DS1/E1/J2 interface and the AAL1 layer. In a typical hardware implementation, the physical interface is supported by a framer chip such as 8260. Its serial output is fed to an

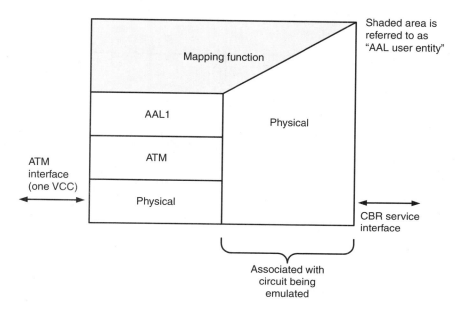

Figure 7.8 UCES interworking function.

AAL1-SAR chip such as Siemens IWE8, which provides a Utopia interface to the ATM network.

7.4.2   Cell Format for Unstructured CES Data Transfer

In the unstructured CES service, the bits entering the interworking function through the serial interface are a serial bit stream, with the framing bits considered part of the data stream. Octet boundaries in the T1/E1/J2 stream lose their relevance when converted to bytes in the ATM cell, but the bit ordering is preserved.

As shown in Figure 7.9, the bit stream is allowed to fill the cell-payload bytes starting from the MSB of the first byte and continuing to the next higher byte with MSB first. Approximately 376 contiguous bits of the incoming stream will fill the 47-byte payload of the SAR-PDU.

7.4.3   Implementation Issues for UCES

Clocking methods for the UCES can be either synchronous or asynchronous. Either asynchronous method, SRTS, or adaptive technique can be used. It is important, however, that both ends of the CES session are

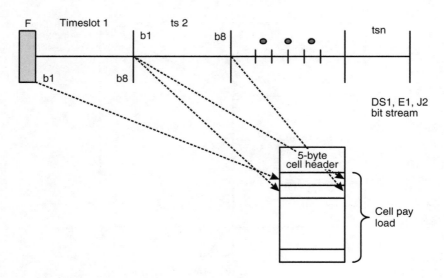

Figure 7.9 UCES cell format/bit ordering.

configured to use the same clock recovery technique. Also, if either end of the CES session is connected to a public T1/E1 or J2 service, the public network operator may mandate the use of the SRTS method to comply with the strict jitter-and-wander requirements demanded by that public network.

Transporting alarm conditions received at the T1/E1/J2 interface, through the ATM network to the far end, is an important requirement for the interworking function. For example, the IWF must detect the loss of signal (LOS) and send AIS cells to the output interface, as shown in Figure 7.10(a).

At the far-end reassembly function, if the buffer underflows, the IWF is expected to insert an all-1s AIS pattern, as shown in Figure 7.10(b). The ATM Forum specifications recommend a period of 2 to 3 seconds of the buffer starvation condition before declaring a red alarm.

7.5   Interworking for Structured CES

Structured CES (SCES) service emulates fractional T1/E1/J2 services, generally termed $N \times 64$–Kbps services. T1, E1, and J2 have, respectively, 24, 32, and 96 slots, each of 64-Kbps data rate. The structured CES services carries multiple slots from one service interface to another.

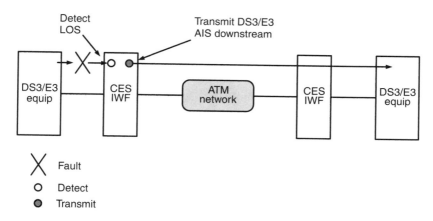

Figure 7.10(a) Unstructured DS3/E3 service interface fault indication.

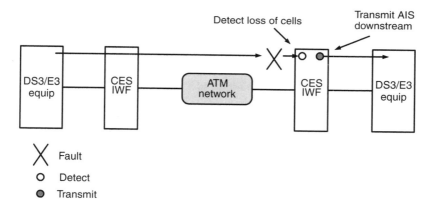

Figure 7.10(b) Unstructured DS3/E3 alarm indication in UCES.

7.5.1 SCES Protocol Stack

Figure 7.11 illustrates the protocol stack organization of the interworking function supporting SCES. The shaded section represents the physical interface side of the circuit being emulated and serves as the user entity for the ATM network side stack supporting the AAL1 adaptation layer.

Since $N \times 64$ service is configured with a fraction of time slots available in the service interface, multiple independent emulated circuits can be supported on a shared service interface, as shown in Figure 7.11. The IWF

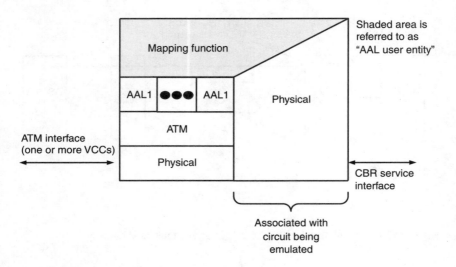

Figure 7.11 CES stack: DS1/E1/J2 structured service interworking function.

provides the mechanism to transfer bits between the physical interface and the multiple AAL1 layer entities, each representing an emulated circuit.

The ATM layer in the protocol stack multiplexes outgoing traffic of multiple VCCs into one ATM cell stream, demultiplexes incoming stream into multiple VCC streams, and presents each stream to the associated AAL1 entity above it.

Similarly, the AAL1 entity for each VCC performs segmentation in the outgoing direction and reassembly in the incoming direction.

The service interworking mapping function above the AAL layer links the stream I/O from the AAL layer to the associated time-slots in the physical layer of the other side of the stack that represents the service interface T1, E1, or J2.

7.5.2 Cell Format for SCES Without CAS

The SDT method described earlier is used as the mechanism to transport data. There are two cell format variations: with CAS and without.

In SCES without CAS, fixed-size blocks of an integral number of data bytes are formatted to fit into ATM cells.

With an $N \times 64$ service such as T1/E1 or J2, the fixed-sized blocks will contain N bytes each. In the specific case where $N = 1$ (for example, a fractional T1 service involving a single DS0), a block will be one byte, and bytes are packed 47 bytes per cell, all in a non-P format (i.e., there are no pointer bytes in any of the cells).

When $N > 1$, N bytes accumulated from each frame will constitute the fixed-size block. The P format will be used, with a cell-coding that satisfies the following rules:

1. The very first cell will be a P format type cell, with the pointer byte set to 0, and the fixed-data block of N bytes starting as the next byte to the pointer byte.

2. Fixed-sized blocks of N bytes will be packed, 46-bytes in this first cell, continuing to fill, with subsequent N bytes, to fill the next seven cells in the sequence (i.e., only one P format cell is allowed in 0 to 7 numbered sequence to satisfy clocking requirements).

In the first possible occurrence during the next sequence of 0 to 7, when a new block begins, the first even-numbered cell will be constructed as a P format cell, with the pointer value pointing to the start of the new block.

The P format cells have even sequence numbers and will have the CSI bit set to 1. Figure 7.11 details a sample set of cells carrying a CES no-CAS service for a fractional T1 setup with $N = 7$.

7.5.3 Cell Format for SCES with CAS

In the CES service with CAS, the format and the coding rules are similar to no-CAS, but the signaling bits are extracted and appended at the end of the fixed-block of data bytes. That is, the overall fixed-size structure will consist of a payload substructure, followed by a signaling substructure.

For CES with CAS, the payload substructure size will include the N-bytes for the total multiframes, as defined for each fractional $N \times 64$ service:

$N \times 64$ service: T1 ESF multiframe $= 24N \times 24$ bytes

$N \times 64$ service: T1 SF multiframe $= 24N \times 24$ bytes

$N \times 64$ service: E1 multiframe $= 16N \times 16$ bytes

$N \times 64$ service: J2 multiframe $= 8N \times 8$ bytes

The signaling substructure byte formats for these four cases are illustrated in Figures 7.12(a–c).

For the T1-ESF case, for each of the time slots the 8th bit of the byte from 6th, 12th, 18th, and 24th are the signaling bits for each time slot.

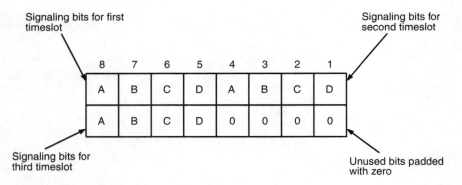

Figure 7.12(a)   DS1/ESF and E1 signaling substructure.

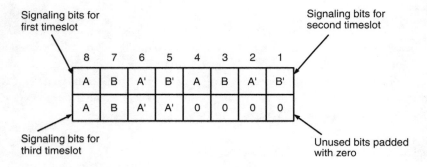

Figure 7.12(b)   DS1/SF signaling substructure.

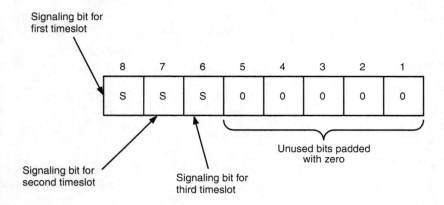

Figure 7.12(c)   J2 signaling substructure.

These bits are extracted and packed to form the signaling substructure, as shown in Figure 7.12(a). For the T1-ESF format, a byte is needed to represent the CAS bits for every two time slots in the fractional service. For example, if fractional T1 service is made up of 17 DS0 time slots, the signaling substructure will be made up of nine octets, the last octet carrying the ABCD bits of the 17th time slot in the upper nibble, with the lower nibble of the 9th octet unused.

For the T1-SF case, data from two adjacent multiframes is sent in a single fixed-sized block. Bits from the first frame, extracted from the 6th and 12th bytes from the 12-frame multiframe, are stored as A and B, and corresponding bits from the second frame are shown as A′ and B′ in Figure 7.12(b).

For E1, a multiframe consists of 16 frames. Time slot 16 of each frame is reserved for the signaling bits. A zero bit pattern in time slot 16 indicates that CAS transmission will start in the following frame. If we denote the zero-pattern frame as CAS begin frame (CBF), then CBF + 1 frame contains the signaling bits for time slot 1 and 17; CBF + 2 frame contains signaling bits for 2 and 17; and CBF + 15 frame contains signaling bits for time slots 15 and 31, as illustrated in Figure 7.12(a).

J2 uses only a single signaling bit per slot in each superframe and, hence, signaling bits are packed for 8-slots/byte as part of the signaling substructure, as shown in Figure 7.12(c).

In these three cases, the signaling bits may be present in the payload substructure, in addition to the signaling substructure, but the destination IWF is expected to use the signaling substructure bits to reconstruct the bit stream that it will then transmit further downstream.

For CES with CAS, the cell formatting rules are the same as that of CES without CAS, as outlined in Section 7.4.3, if the fixed-size block length is considered as two fixed-size subblocks, the payload substructure and the signaling substructure. An example cell format for a 24-time slot T1 with ESF format is shown in Figure 7.13.

7.5.4 Implementation Issues for SCES

Interworking functions at either end of the network supporting structured CES services must be served by clocking arrangements traceable to a PRS. A typical arrangement identified in the ATM Forum specification [1] is that a PRS traceable source supplies timing to the physical layers of the ATM links between the IWF and the ATM network. Timing might be introduced to

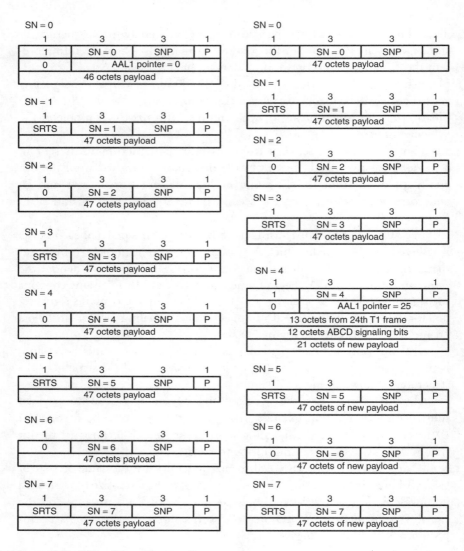

Figure 7.13 CES cell format expansion.

the ATM network by connecting each ATM switch to a central office clock. Each CES IWF then receives its timing from its ATM interface.

As with UCES, transporting alarm conditions received at the T1/E1/J2 interface, through the ATM network, to the far end is an important requirement for the interworking function. For example, when LOS, out-of-frame, or AIS occur, the IWF applies trunk conditioning in the downstream direction, as shown in Figure 7.14(a). It emits cells at the nominal rate but sets the payload to indicate idle or out-of-service condition.

At the far-end reassembly function, if the buffer underflows, the IWF is expected to insert an all-1s AIS pattern, as shown in Figure 7.14(b). The ATM Forum specifications recommend a period of 2 to 3 seconds of the buffer starvation condition before declaring a red alarm.

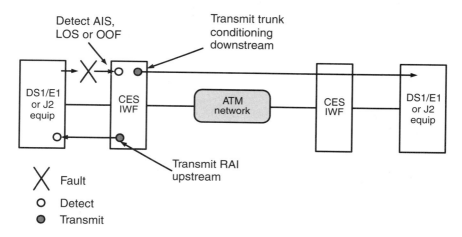

Figure 7.14(a) *N*x64 service interface fault indication.

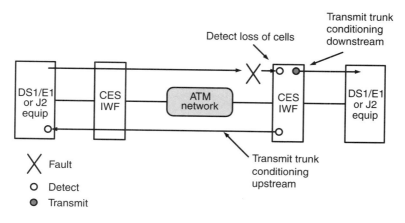

Figure 7.14(b) Virtual channel fault indication.

7.6 Summary

This chapter focused on the implementation details of different CES configurations. Chip-sets available in the market today currently allow system designers to provide the CES functionality totally in hardware, except for the initial configuration of the services. The AAL1 SAR chips, together with the service interface T1/E1/J2 framer circuits, jointly provide all necessary functions. SAR chips provide Utopia interfaces to the ATM part of the hardware that ultimately provides the interface to a public ATM network.

References

[1] The ATM Forum Technical Committee, "Circuit Emulation Service Interoperability Specification Version 2.0," af-vtoa-0078.000, January 1997.

[2] ITU-T Recommendation I.363, "B-ISDN ATM Adaptation Layer (AAL) Specification," March 1993.

[3] ITU-T Recommendation G.702, "Digital Hierarchy Bit Rates," June 1990.

[4] ITU-T Recommendation G.823, "The Control of Jitter and Wander Within Digital Networks Which Are Based on the 2,048 Kbps Hierarchy," March 2000 (Revision 3).

[5] ITU-T Recommendation G.824, "The Control of Jitter and Wander Within Digital Networks Which Are Based on the 1,544 Kbps Hierarchy," March 2000.

8

Traffic Management in ATM Networks

8.1   Introduction

In previous sections of this book, we have covered functional aspects of inter-working between ATM and other dominant access and network protocols. IP technology, with its widespread use in the World Wide Web and as the chosen technology to carry voice in the emerging convergence of PSTN with packet networks, was used as the main protocol for illustrating interworking with ATM.

   This chapter lays the framework for understanding another important aspect of ATM networking: traffic management. When the same network supports different types of services, it becomes necessary for network elements to differentiate between service types and support mechanisms that provide a predictable and consistent level of quality appropriate to each service. Whereas IP networks are traditionally designed to provide best-effort service, ATM networks must support multiple service types simultaneously, requiring a framework for precise traffic-management techniques to ensure that subscribers receive levels of quality expected of their service class subscriptions.

   In this chapter, traffic-management concepts are illustrated with a view of understanding design challenges that arise when provisioning services with specific QoS requirements in networks that contain:

- A wireless segment;
- Interworking between multiple protocols with different QoS characteristics.

The presence of a wireless segment in a service connection makes guaranteeing a defined level of service more complex. In a TDM-based wireless protocol, the cell-arrival pattern at the transmitting end is modified by the slot availability in the air-interface, producing a different cell traffic pattern at the receiving end. The cells are also likely to experience increased delay. These factors affect the QoS characteristics of a session.

The information in this chapter is also helpful to understand the challenge of guaranteeing QoS in future networks where an end-to-end service session includes segments carried by ATM and different high-speed network protocols. In addition, the material illustrates how network mechanisms are provided to enforce the QoS requirements at the network boundary and inside networks, which include wireless segments.

ATM traffic-management concepts can be used to address these two factors, based on ITU-T Recommendations I.371 and I.356 and ATM Forum Traffic Management Specification TM4.1.

8.2   Traffic-Management Overview

Network users need bandwidth at the lowest cost possible. Network service providers supporting different service types need to build and manage their network so that bandwidth needs are supported with minimum resources. Services differ in terms of bandwidth requirements, real-time needs (tolerance for delays), and ability to survive loss of frames or cells at the network layer level. The challenge facing network designers is to construct a network with features that support the varying service needs and, at the same time, make use of the flexibility inherent in service multiplexing to obtain effective utilization of available bandwidth. The ATM traffic-management framework provides the tools necessary to achieve these objectives.

The basic components in the implementation of ATM traffic management are shown in Figure 8.1. The application indicates its expected level of service and characterizes the behavior of its source traffic using a traffic contract. It negotiates the parameters of the contract with the network and comes to an agreement. The main element of the traffic contract is a service class that specifies the type of service required for the connection. The service

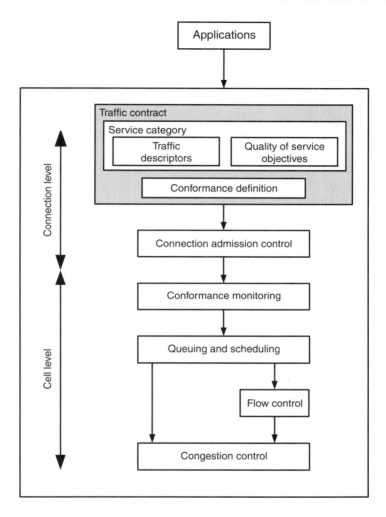

Figure 8.1 Component functions of traffic management. (*From:* Giroux/Ganti, Prentice Hall, 1999.)

class broadly defines the expected QoS of the connection, and the expected behavior of the source traffic represented by traffic parameters. A CAC algorithm then determines if the network can safely support the negotiated traffic contract without impacting the QoSs contracted for existing connections. These components of traffic management, as shown in the diagram, are executed during the connection setup phase, before the connection transports data.

Other components of traffic management take place at the cell level during the flow of cell stream in a service session. From the traffic contract, the boundaries for the behavior of cells are parameterized in the form of a conformance definition. Generally, within ATM-cell processing hardware, standard algorithms can be configured to police the cell stream for conformance to this definition. Policing is performed at a cell stream's network entry point (ingress) to identify illegal cells and take appropriate action so that other service streams are not affected by rogue cells that exceed conformance parameters.

On intermediate links within a service provider network, different services are multiplexed and demultiplexed to achieve gains due to statistical multiplexing. Multiplexing involves queuing of cell streams at different transmit points in the network and implementing scheduling algorithms that recognize the QoS needs of different service classes. Since flow characteristics of cell streams are altered by the actions taken at intermediate nodes in the network, at the cell stream's exit point (egress), the network may attempt to shape the cell stream so that flow characteristics are brought closer to the agreed behavior specified by the traffic contract.

Congestion at queuing points is still possible if bursts from different streams overlap; congestion handling schemes must therefore be implemented to discard cells from different traffic classes in accordance with different QoS needs. Queuing, scheduling, and congestion handling are important components of traffic management, but they are not the primary focus of this book and will not be discussed.

8.3 Traffic Contract

In an ATM network, the traffic contract provides the basis for supporting different types of traffic. The user effectively obtains agreement from the service provider that the network will provide the user session(s) a level of service described in the traffic contract. A traffic contract for a service contains the following key components:

- Specified service category in the ATM service architecture.
- Required QoS, defined by a set of parameters.
- Traffic characteristics of the connection.
- Conformance definition.

In a fully developed ATM network that supports SVCs, a traffic contract is negotiated using ATM signaling. In limited ATM network implementations, where only PVCs can be configured, traffic contract components are specified by the network management system.

The components of traffic contracts are described in further detail in the following sections.

8.3.1 ATM Service Architecture

Six service categories can be supported by the architecture for services provided at the ATM layer:

- *CBR:* constant bit rate service;

- *rt-VBR:* real-time variable bit rate service;

- *nrt-VBR:* non-real-time variable bit rate service;

- *UBR:* unspecified bit rate service;

- *ABR:* available bit rate service;

- *GFR:* guaranteed frame rate service.

CBR and rt-VBR services fall into the category of real-time services, where applications are very sensitive to variations in end-to-end transmission delay. For these services, bandwidth is allocated during connection establishment, whereas for UBR, ABR, and GFR services, bandwidth is dynamically allocated on demand during the connection's lifetime. Among the bandwidth-on-demand services, ABR and GFR guarantee a minimum amount of bandwidth. These bandwidth-on-demand services need robust flow control mechanisms to accommodate the variation in data rates introduced by statistical availability of bandwidth. In ABR, tight flow control procedures are implemented at the ATM layer level, whereas for UBR and GFR, higher application layers at the end systems perform flow control.

8.3.1.1 CBR Service

CBR service mainly supports real-time applications that require a fixed amount of continuously available bandwidth, defined by a parameter PCR. These applications demand tightly constrained delay variations; the source can emit cells at PCR at anytime or be silent. Interactive video and audio, video and audio broadcasts (e.g., distributed classroom and audio feeds) and video and audio retrieval (e.g., VOD and audio library), and CES are typical applications that fall into this category. The curve labeled CBR in Figure 8.2 shows the typical cell rate profile for a CBR application. The parameter CDVT reflects the allowed temporary instantaneous peaks and valleys in the traffic profile. Total end-to-end delay is limited by a parameter maxCTD, and cells violating this rule are assumed to be of low value to the application, eligible for discard.

8.3.1.2   rt-VBR Service

In this class of real-time applications, the source emits cells at a rate that varies with time, characterized by SCR, a value reflecting the average cell rate. The real-time constraints are similar to those in the CBR category, with the parameters PCR and CDVT also applicable to this flow, as illustrated in Figure 8.2 by the curve labeled rt-VBR. The source is permitted to emit bursts exceeding SCR (but below PCR), up to the number of bits defined by MBS. Real-time applications that can benefit from statistical multiplexing and recover from small loss of cells can use this service. Voice and variable-bit-rate video are representative applications of this class.

8.3.1.3   nrt-VBR Service

The traffic profile for nrt-VBR applications is similar to that for the rt-VBR (i.e., defined by PCR, SCR, CDVT, and MBS), but CTD constraints (i.e., MaxCTD) do not apply. CDVT and MBS values are likely to be large,

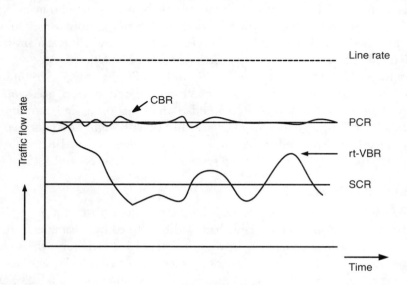

Figure 8.2 CBR and rt-VBR traffic rate profiles.

indicating high tolerance to real-time aspects of cell delays. Also, these applications are characterized as those demanding low CLR. Applications in the nrt-VBR category include packet data sessions, terminal sessions (e.g., airline and banking transactions), and file transfers.

8.3.1.4   UBR Service

As a non-real-time service, UBR resembles the best-effort service associated with IP nets. This service has no constraints on absolute end-to-end delay values or delay variations. UBR does not provide guarantees on throughput or any QoS parameters. E-mail, file transfers, and non-mission-critical traffic fall into this service category.

8.3.1.5   ABR Service

ABR service is intended for non-real-time services where a minimum amount of bandwidth, or minimum cell rate (MCR), is guaranteed, and the peak rate of emission of cells is limited to a defined PCR value. Flow control is activated independently in both directions. Figure 8.3 shows an example where the source sends resource management (RM) cells in the forward direction, which are then sent around at the destination as reverse-RM cells. These cells carry feedback information related to the availability of bandwidth in the different network elements in this path, allowing the source to modify its

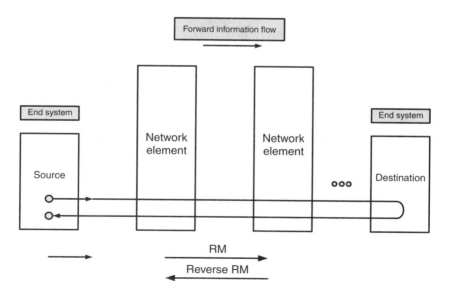

Figure 8.3 Example of ABR control loop.

source traffic rate accordingly. Network elements may also generate their own reverse-RM cells, indicating the level of congestion. Network elements can also set EFCI bit in the forward cell headers. When these cells reach the destination node, the destination can generate reverse-RM cells to inform the source node of the congestion. ABR services do not have constraints in end-to-end delay or delay variation requirements. This control-loop ATM layer feedback mechanism helps to minimize cell losses and allows the service to make maximum use of dynamically varying spare bandwidth. Typical applications that can use ABR service are LAN interconnection, database archival, and Web browsing.

8.3.1.6   GFR Service

GFR service is another service to support non-real-time applications. During connection establishment, MCR (which can be zero) PCR, MBS, and maximum frame size (MFS) parameters are specified. No feedback mechanism is enforced; therefore, traffic beyond MCR will only be delivered if network resources are available. Data is discarded at the AAL5-PDU level, thereby avoiding random discarding of cells without reference to any frame boundary. The user can send frames either marked (CLP = 1) or unmarked (CLP = 0), where all cells in a frame have the same CLP value. The MCR guarantee only applies to unmarked frames.

Similar to other non-real-time services, GFR is not constrained by any requirements in delay or variations in delay. Frame Relay over ATM is a typical application that is likely to use this service.

These service categories define the boundaries of QoS parameters and traffic descriptors, as will be seen in the following sections.

8.3.2   QoS

A set of parameters characterizing the end-to-end performance at the ATM layer of a connection is used as the measure of QoS for that connection. ITU Recommendation I.356 refers to these as network performance parameters. Different sets of QoS parameters have to be negotiated by the network and the end systems for each direction of data flow.

UNI Signaling 4.0 and PNNI 1.0 standards provide signaling mechanisms to negotiate QoS in an ATM network that supports SVCs. In PVC-only systems, the QoS parameters are defined statically by the network management system during service provisioning.

Three negotiable and three nonnegotiable parameters form the QoS parameter set, as defined in Table 8.1 and in the following sections.

Table 8.1
Negotiable and Nonnegotiable Parameters

Nonnegotiable Parameters	Negotiable Parameters
SECBR: severely errored cell block ratio	P2P-CDV: peak-to-peak cell delay variation
CER: cell error ratio	MaxCTD: maximum cell transfer delay
CMR: cell misinsertion rate	CLR: cell loss ratio

8.3.2.1 Nonnegotiable QoS Parameters

These nonnegotiable parameters reflect the transmission system efficiency and measure the errors in the service session caused by transmission problems. A block is said to be severely errored if a number (M) of cells in the block (defined by I.610 as a number of information cells between two OAM cells) are in error. The severely errored cell block ratio (SECBR) is then the ratio between the number of blocks in error and total blocks transmitted. Identifying severely errored cell blocks and not including these cells in other error parameters avoids the skewing of quality parameters by a burst of errored cell blocks. Misinserted cells are those with a bad but undetected cell header being mapped into a valid VP/VC session; cell misinsertion rate (CMR) is usually in the range of a few per day. CER measures the ratio of cells with bad headers that cannot be recovered (using error correcting algorithms) against total cells transferred.

8.3.2.2 Negotiable QoS Parameters

The negotiable parameter CLR measures the number of lost cells, cells with bad headers, and cells with corrupted payload, against total transmitted cells. Buffer overflows are likely to be the main cause of cell loss; therefore, the CLR value and the effectiveness of traffic-management schemes are closely linked. Applications that negotiate nonzero CLR values should either be able to function with a low rate of cell loss or have higher protocol layers that can recover from cell loss. An application that subscribes to nrt-VBR is likely to negotiate and expect low CLR values when compared to applications that request bandwidth-on-demand services.

P2P-CDV and maxCTD are the most significant negotiable QoS parameters, as they differentiate between real-time and non-real-time services. Both parameters are based on the end-to-end delay characteristics of a session supported by one or more private/public ATM networks. A reference network configuration for QoS measurement is shown in Figure 8.4.

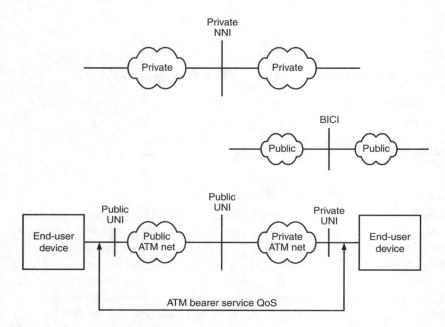

Figure 8.4 Typical QoS reference configurations.

The CTD is the aggregation of delays introduced by each node in the session. These include queuing, switching, and processing delays, as shown in Figures 8.5(a, b). If a TDM-based wireless protocol forms a segment of the link, the time for cells to wait until they encounter a valid air-slot will also add to the end-to-end delay, as shown in Figure 8.5(c).

The relationship between the parameters peak-to-peak cell delay variation (P2P-CDV) and maxCTD can be seen in a typical probability density function of CTD showing the fixed delay and the variable delay in an end-to-end session, as shown in Figure 8.6.

8.3.3   Traffic Descriptors

Source traffic descriptors describe the expected characteristics of the traffic entering the network and are defined during an ATM connection configuration time as part of the traffic contract. The source descriptor set consists of one or more of the following parameters:

- PCR;
- SCR and MBS;

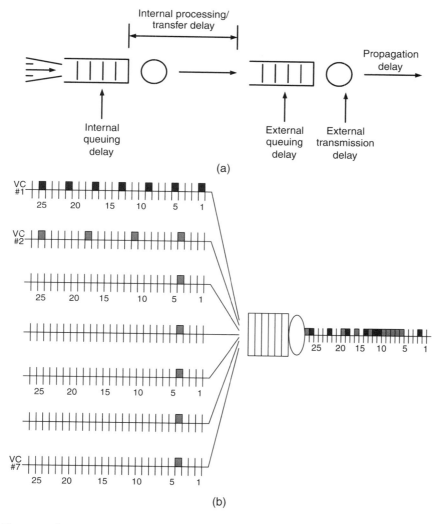

(a)

(b)

Figures 8.5(a, b) Factors causing cell delay variation.

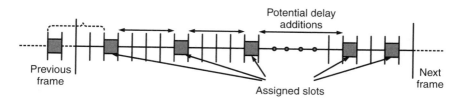

Figure 8.5(c) Delay due to TDM slot availability.

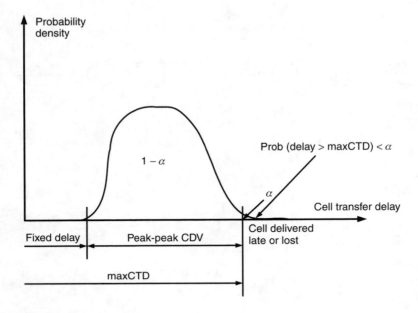

Figure 8.6 Illustration of CTD.

- MCR;

- MFS.

In specifying traffic descriptors and defining traffic conformance, the CLP bit in the ATM cell header plays an important role. Descriptors either refer to flows with CLP = 0 or to aggregate flows where CLP is 0 or 1. Similarly, conformance definitions that refer to CLP have to be implemented according to TM4.1 to ensure interoperability.

CDVT is also added during the connection setup. This parameter is used in the conformance checking of traffic flow and is always greater than the maxCTD.

When both CLP = 0 + 1 is specified, PCR is always specified with a value that must be less than the line rate. Figure 8.7(a) shows a typical cell flow with the PCR value of half the line rate. For example, if a CES of bandwidth 4xDS0 is supported by a service, the PCR is defined as 683 cells/s, or, alternately, if a Frame Relay circuit is carried by a full T1, the peak cell rate is defined as 5,052 cells/s, both reflecting the physical limit of line capacity. In the Frame Relay example, multiple DLCIs can be supported in the same physical circuit, and PCR values that are less than the maximum line rate can be given for individual DLCIs, but the sum of all PCRs can be more than the full T1 capacity. The key point to note here is that the sourcing DLCIs cannot overload the system with a traffic rate of more than T1 (5,052 cells/s) because the aggregate is limited by the line capacity.

The SCR traffic descriptor specification is also accompanied by an MBS. The average cell rate of a service session without exceeding the PCR is denoted by SCR. MBS defines the number of cells that can be sent in a burst at the PCR without violating the conformance rule. In describing the flow in SCR terms, the CLP value becomes relevant. Two types of flows having the same SCR and MBS specifications can differ based on whether only CLP = 0 cells or both CLP = 0 + 1 cells are considered for flow conformance. Figure 8.7(b) shows the case where both CLP = 0 + 1 cells are considered, and Figure 8.7(c) shows the case when only CLP = 0 cells are considered.

In the same Frame Relay example, the sum of all SCRs specified for the DLCIs on a T1 line cannot be more than the maximum cell rate supported by physical T1 capacity.

MCR is used by bandwidth-on-demand services ABR and GFR to guarantee at least this minimum amount of bandwidth, thereby ensuring a minimum amount throughput for the connection if bandwidth becomes scarce in the network.

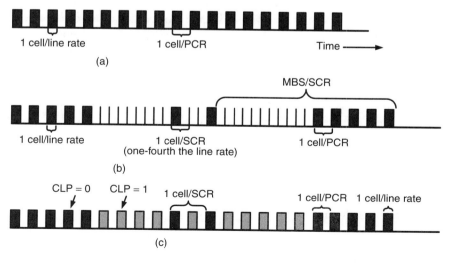

Figures 8.7(a–c) Cell flow examples illustrating line rate/PCR/SCR/MBS. (*From: Giroux/Ganti, Prentice Hall, 1999.*)

MFS is applicable to the GFR service; it specifies the maximum frame size for AAL PDUs. The network need not meet the QoS objectives for frames larger than MFS.

8.4 Traffic Conformance

After a traffic contract is negotiated, the connection control algorithm determines the resources required to meet the QoS needs and allows or disallows the connection to be set up based on the availability of system/bandwidth resources. However, the data stream in the admitted connection may exhibit behavior that may need further network action for the following reasons:

- At the input to the network (ingress point), the source stream may not adhere to the established traffic descriptors and may begin to affect the QoS of other data streams active in the network.
- Due to internal queuing, rescheduling, or the delay added by a wireless link within the network, traffic flow characteristics may change to the extent of violating the established traffic parameters when the cells exit (egress) the network.

The network therefore needs to monitor the behavior characteristics of the cell stream at the ingress to identify cells not conforming to the agreed traffic contract. These cells are either dropped or tagged based on the contract agreement. This procedure is known as *policing*. Traffic contract conformance is determined by a GCRA defined in ATM Forum specifications and explained in Section 8.4.1.

At the egress, the network attempts to smooth the traffic flow to make the cell stream conform to the established traffic descriptors as much possible so that, at the entry to upstream network, the policer does not act harshly on the traffic stream. This function at the egress is known as *shaping*, as explained in Section 8.4.2.

8.4.1 Traffic Policing

Policing functions act on traffic streams entering a network (at the ingress point) to ensure that the incoming traffic conforms to the agreed traffic contract. Since network bandwidth is allocated assuming conforming cells, policing guarantees QoS to conforming cells in cell streams.

Policing functions are generally nonintrusive and leave the cell flow characteristics unchanged, except for dropping or tagging nonconforming cells. Altered cell flow patterns due to delays caused by queuing and congestion within a network can be impacted by policing functions. Traffic streams are policed either at the ingress point of a network or after a shaping function.

Although implementation of policing functions is not standardized, stricter algorithms (i.e., which find more nonconforming cells) than the GCRA should not be used. A more forgiving algorithm is acceptable but would require allocation of more resources to guarantee the same levels of QoS.

In practical implementation of policers, generally use fixed increments of traffic rates to police traffic streams. During connection setup, the next higher level of supported traffic rate is generally chosen as the policing traffic rate level. Clearly, the efficient use of chargeable bandwidth depends on the granularity of traffic rates; the finer the granularity, the greater the efficiency with which bandwidth can be allocated to user sessions.

The conformance algorithm with a PCR and CDVT can be applied as a policer to CBR traffic. For policing VBR connections, however, a dual version of the GCRA algorithm—one to monitor the PCR and the allowable CDVT and the other to monitor the MBS specification in conjunction with the SCR—is required. GFR policing can be similar to that applied to CBR traffic augmented with an additional check on the MFS. For UBR, policing similar to that applied to CBR can be used to keep track of the PCR with a large CDVT to accommodate larger queuing delays and jitter in the upstream nodes.

8.4.2 Traffic Shaping

Traffic shaping can be used to modify a traffic flow to conform to contracted traffic descriptors. Shaping is employed in a network under the following scenarios:

- Devices generating the traffic (e.g., an application running in a PC) are unlikely to generate traffic that conforms to defined traffic descriptors such as PCR and CDVT. This illustrates the need to employ shaping at the traffic source point so that the traffic that enters a network conforms to the traffic descriptors contracted with the network.

- Multiplexers of traffic and traffic aggregation points, where multiple VCs are combined into a single VP, generally introduce jitter and can produce varying traffic flow characteristics from shaped traffic input. When multiple VBR or CBR traffic is aggregated, one can see the need to employ shaping to obtain efficient utilization of bandwidth. For example, when five VBR connections with PCR = 0.1, SCR = 0.05, and MBS = 10 are combined, one can assume a resulting SCR of 0.25 and worst-case values for PCR of 1.0 and MBS of 50 cells. Since it unlikely that the worst-case traffic profiles of all the VCs will coincide with time, this assumption will result in underutilization of bandwidth. Traffic shaping will ensure traffic conformity for a VP when the dynamic characteristics and the number of VCs change.

- Traffic shaping may be necessary at the egress of a network before delivering the traffic to the destination customer node or before allowing traffic to enter another public network. The first case is a value-added service that will reduce the buffering needs in customer end-nodes, and the second case will reduce possibilities of traffic nonconformity with the contracted traffic descriptors at the upstream public network that will be policing the entering traffic.

Conformant traffic streams can be produced by different traffic shaping algorithms. Popular ones follow the reverse form of the leaky bucket algorithm. Since the primary goal of a shaper is to minimize the CDV, the CDVT parameter in the traffic contract is never deliberately used to send cells back-to-back such that they just avoid nonconformance. However, shapers will likely encounter collisions of cells in the multiple incoming streams and a nonzero CDVT to accommodate cells that are delayed due to collisions.

8.4.3   Generic Cell Rate Algorithm

In policing and shaping implementations, a method known as the generic cell rate algorithm (GCRA) is generally used on a cell-by-cell basis to test conformance of traffic with respect to the established traffic contract.

The ATM specifications formally define usage parameter control (UPC) as the set of actions taken by the network to monitor traffic and enforce the traffic contract at the UNI. The specification does not mandate the use of GCRA as part of UPC, but instead requires UPC to use any algorithm that supports the QoS objectives of a compliant connection. Several algorithms are discussed in Sections 8.4.3.1 to 8.4.3.5.

8.4.3.1 Virtual Scheduling and Leaky Bucket Algorithms

The GCRA can be a *virtual scheduling algorithm* or a *continuous-state leaky bucket algorithm*, as shown in Figure 8.8.

Given an average cell rate and the associated tolerance in cell delay variation, the GCRA identifies nonconforming cells. These two parameters, which define the operation of GCRA, are expressed as follows:

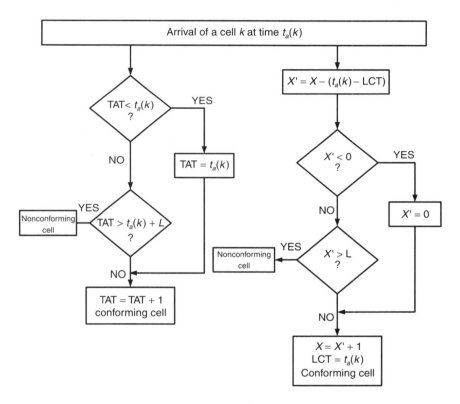

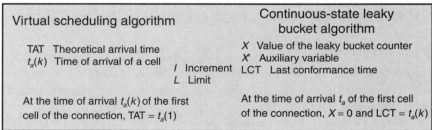

Figure 8.8 Equivalent versions of the generic cell rate algorithm.

- *Increment (I)*—average interarrival time of cells for the given average cell rate; a reference unit of time is chosen to express *I*.
- *Limit (L)*—CDVT expressed in the same time unit.

In virtual scheduling, the algorithm keeps track of the theoretical arrival time (TAT) of a cell. If the difference between TAT and the actual arrival time t_a is greater than L, then the cell is declared nonconformant. If the cell arrives later than TAT (i.e., $t_a >$ TAT), TAT is set to t_a. TAT for the subsequent cell is derived by incrementing the current TAT by *I*. At the start of the algorithm, TAT is initialized to the arrival time of the first cell. The algorithm shown on the left of Figure 8.8 then identifies subsequent cells as conforming or nonconforming.

The algorithm can also be described as a *leaky bucket*. In this case, a bucket is filled with *I* units on the arrival of each conforming cell. The bucket is assumed to drain at the rate of 1 per unit-time. If, at a cell arrival, the content of the bucket is less than or equal to *L*, then the cell is conforming. The upper and lower bounds of the bucket are $L + I$ and 0, respectively. If the content of the bucket reaches zero, draining (decrements) stops until the bucket level has a positive value at the arrival of the next cell. The right half of Figure 8.8 illustrates the operation of the algorithm, where at the arrival of the first cell, the contents of the bucket are set to zero.

For all cell arrival patterns, the behavior of both the virtual scheduling and leaky bucket algorithms are equivalent in identifying nonconforming cells.

8.4.3.2   Conformance for PCR Flow

For a CBR service, where only the PCR rate is involved, a single leaky bucket algorithm or virtual scheduling can be applied to test conformance. Theoretically, cells should arrive with 1/PCR (denoted as T_{pcr}) interarrival time. However, to account for the jitter of cells, CDVT is used as a tolerance factor. These two parameters serve as the GCRA parameters ($I = T_{pcr}$, $L =$ CDVT) to test for conformance of PCR flows.

8.4.3.3   Cell Clumping

A closer look at the impact of cell clumping on PCR conformance provides insight into how the same GCRA algorithm can be used to test conformance for SCR flows.

Figures 8.9(a–d) illustrate some scenarios of cell clumping. One needs to define another parameter *D* that describes the minimum interarrival time

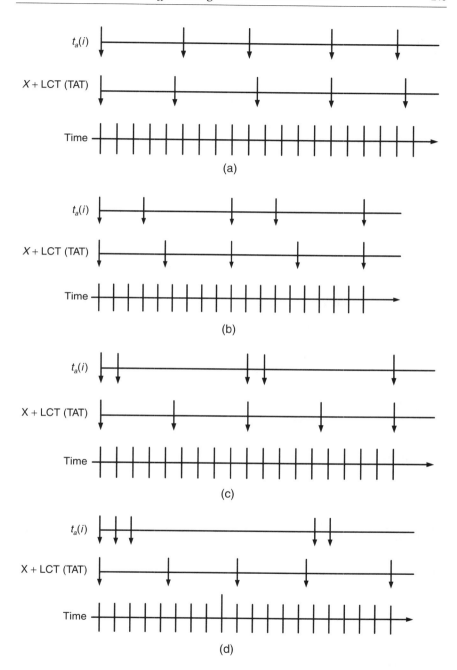

Figures 8.9(a–d) Conformance in cell clumping scenarios.

determined by the line rate. The PCR and the interarrival time can be expressed as a multiple of D and assume the value $T = 4.5D$. In Figure 8.9(a–d), conforming cell arrival scenarios are shown for $L = 0.5D$, $1.5D$, $3.5D$ and $7D$.

In Figure 8.10 shows a more generic depiction, where cells are clumped in successive slots represented by the maximum line rate. Here, the condition for conformance can be written as follows:

$$(n - 1)I - (n - 1)D < L$$

$$(n - 1) < L / (I - D)$$

$$N < 1 + (L / (I - D)) \qquad (8.1)$$

This inequality reveals that the upperbound on clumping is a measure of (or directly related to) the CDVT (or limit L in the GCRA).

8.4.3.4 Conformance for SCR Flow

Parameters for defining the SCR flow in the GCRA algorithm can be easily understood by referring to the cell clumping model shown in Section 8.4.3.3 and the periodic on-off source representing the worst-case traffic pattern that will be compliant with the SCR flow traffic descriptors.

If the (MBS is given, then we may derive the equivalent value for L (in the SCR flow this is called BT) using the cell clumping formula shown in (8.1). Since PCR flow will restrict the maximum cell rate to the value given by PCR, the minimum interarrival time for the SCR flow is determined by 1/PCR. In a worst-case scenario, MBS represents the maximum number of

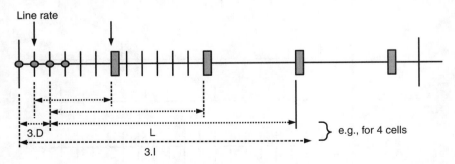

i.e., 3.I – 3.D < L for passing conformance generalizing, (n – 1)I – (n – 1) D < L
Gives max number of clumped cells (at line rate) which pass conformance

Figure 8.10 Inequality for upper-bound on clumping.

cells that can be clumped together without violating the SCR flow conformity. Hence,

$$MBS < 1 + \text{Integer } [BT/(1/SCR - 1/PCR)]$$

that is, $MBS < 1 + \text{Integer } [BT/(T_{scr} - T_{pcr})]$, where the interarrival times for PCR and SCR flows are denoted as

$$T_{scr} = 1/SCR$$

$$T_{pcr} = 1/PCR$$

Also note that, given MBS, PCR, and SCR, the value of BT is not uniquely determined but can be any value in the closed interval,

$$[(MBS-1)(T_{scr} - T_{pcr})] \text{ and } [MBS(T_{scr} - T_{pcr})]$$

However, to ensure consistency, by convention, the minimum possible value is used for BT and is thus defined by

$$BT = (MBS - 1)(T_{scr} - T_{pcr})$$

Similar conclusions can be derived by observing the behavior of the on-off source shown in Figure 8.11. The source emits cells at peak rate PCR for a maximum duration of MBS/PCR. This represents the rule for conformance for an SCR flow, where the number of cells sent at PCR rate cannot exceed MBS. A similar pattern can again begin with a cycle time of MBS/SCR.

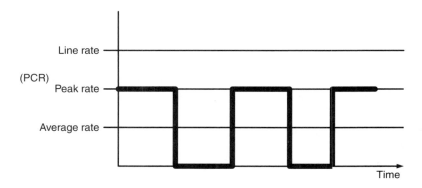

Figure 8.11 Example of on-off behavior to derive GCRA for SCR flow.

One sees that if the SCR flow conformance is not to be violated in a flow that has transmitted at PCR rate MBS number of cells, the next cell cannot be emitted until a lapse of MBS($T_{scr} - T_{pcr}$) time units. This time period represents the higher bound for BT, as described earlier.

An additional consideration is required to define the limit parameter for GCRA to express conformance for SCR flow. Since BT does not take into account allowable jitter of the cells upstream, the value of CVDT is added to the BT value as the limit parameter for the GCRA algorithm. This results in conformance for SCR flow being defined by the same algorithm that is used to check the PCR flow, but with the following parameters:

$$I = T_{scr}$$

$$L = \text{CVDT} + \text{BT}$$

The conformance test is defined as GCRA (T_{scr}, CVDT + BT).

8.4.3.5 Clumping and Wireless Protocols

Air-protocols that support any wireless segment in a network are usually based on TDMA techniques using different forms of modulations for the physical layer. Common among modulation schemes used in wireless systems are QPSK, 16-QAM, and 64-QAM techniques. Quad-forms of these, where four cells are packed in a burst of data, are also quite common.

In an air-protocol with TDM configuration, a TDM frame will be organized into multiple time slots, each slot capable of carrying an rf-burst. In a scheme where a mixed mode of quad-QPSK, quad-16 QAM and quad-64 QAM can be supported, a slot in the densest configuration will carry a burst of quad-64 QAM. Figure 8.12 shows a typical TDM-based wireless configuration.

For efficient operation of the air-protocol, four cells are sent back-to-back in a burst; the average rate will depend on the modulation type chosen. If no shaping is done at the downstream end of the air-protocol, any policing operation will have to accommodate, at the minimum, clumping of four cells. Based on the criteria chosen to allocate multiple slots to accommodate the expected data rate of the TDM stream, additional issues may arise to ensure that the stream passes the policing test at the downstream egress. Specific examples from different ATM interworking scenarios that include a wireless segment will be covered in Chapter 9.

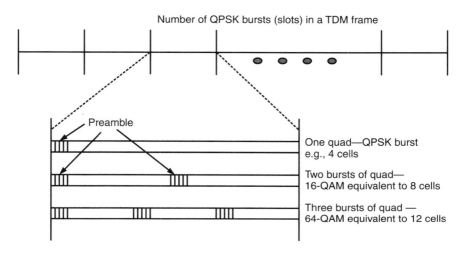

Number of QPSK bursts (slots) in a TDM frame

Preamble

One quad—QPSK burst
e.g., 4 cells

Two bursts of quad—
16-QAM equivalent to 8 cells

Three bursts of quad —
64-QAM equivalent to 12 cells

Figure 8.12 Typical TDM-wireless/air-protocol configuration.

8.4.4 Frame Relay Traffic Profiles and Equivalent ATM Cell Flows

ATM cell flows resulting from a Frame Relay interworking function are governed by several factors, including the following:

- The frame length of Frame Relay PDUs will determine the number of cells when the AAL5 conversion takes place. A CBR Frame Relay input with varying frame lengths will result in a nonconstant ATM cell rate. Table 8.2 and Figure 8.13 illustrate this mapping for a CIR of 1.536 Mbps (T1 rate).

- Frame Relay traffic policing is performed at relatively large time windows (1 sec or more) where there will be a clear demarcation of conforming (Bc) frames and excess frames (Be). In contrast, traffic policing in ATM is carried out at the cell arrival instances and, therefore, finer grained traffic policing is enforced. CDVT and BT values that determine the ATM criteria for cell tagging and dropping must reflect this characteristic.

- The issue related to timeframes used in policing in ATM and Frame Relay is further complicated if a wireless component is introduced for transporting ATM cells carrying Frame Relay data (service or network interworking). In particular, if the wireless protocol is based on TDM, the slot allocations in the TDM frame determine the

Table 8.2
Frame Rate and Cell Rate

Frame Length (Excludes 2-byte DLCI 8-byte AAL5 Trailer)	Frames/s	Cells/Frame	Cells/s
1	19,200	1	19,200
38	5,052	1	5,052
39	4,923	2	9,846
86	2,233	2	4,466
87	2,207	3	6,621
134	1,433	3	4,299
135	1,422	4	5,688
182	1,055	4	4,220
183	1,049	5	5,245
230	835	5	4,175
231	831	6	4,986
278	691	6	4,146
279	688	7	4,816
336	571	7	3,997
337	570	8	4,560
518	371	11	4,081
519	370	12	4,440
1,046	184	22	4,048
1,047	183	23	4,209
2,046	94	43	4,042

clumping characteristics of the ATM cell flow at the receiving hub terminal. The ATM traffic policer at the hub needs to be configured to accommodate the delays introduced by the TDM slot assignment algorithms.

In practical implementations, direct Frame Relay to ATM interworking modifies traffic profiles as cells are generated and transmitted from Frame

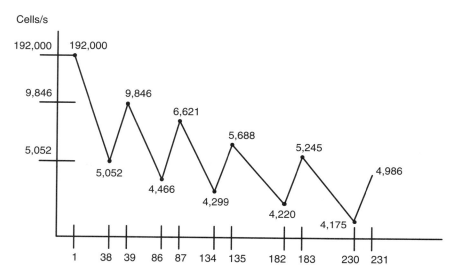

Figure 8.13 Cell rate against frame length for a fixed CIR (Frame Relay) traffic of T1 speeds.

Relay frames. Buffering techniques and policing implementation in Frame Relay software are main contributors to this. These factors are explored with specific reference to selected Frame Relay traffic profiles in direct interworking implementations before the impact of wireless medium are examined.

8.4.4.1 Constant Bit Rate at Bc

Setting aside the variability in the frame sizes, if we assume that the Frame Relay traffic profile can be characterized as CBR traffic at a specified CIR, the two associated traffic profiles are constant. The equivalent SCR for the given CIR (or Bc) can be obtained from a table similar to Table 8.2 but derived for the specific bit rate of CIR.

In practice, it is hard to produce a CBR traffic from a Frame Relay source. One possible scenario is when this rate is defined and limited by the line capacity (i.e., fractional T1, T1, or T3 speeds). Further, if the frame sizes are kept the same, then on the ATM cell flow profile one can expect a constant cell rate.

8.4.4.2 Constant Bit Rate at Bc + Be

Whereas Bc is defined as the maximum amount of data (bits) that a Frame Relay network agrees to transfer over a measurement interval *T*, Be is the uncommitted data (bits) the network will attempt to deliver over the same measurement interval. The excess burst data is generally marked as DE by the Frame Relay policer.

In a typical flow defined by a CBR traffic at Bc + Be rate, as shown in Figure 8.14(a), where Bc and Be are the same, during the first half of the measurement interval the Frame Relay policer will allow the Frame Relay frames as conformant and mark the frames in the second half of the measurement interval as DE. If the rate exceeds Bc + Be, then during the tail end of the interval, the nonconformant frames (i.e., frames whose bits have exceeded Bc + Be during that interval) will be dropped.

The corresponding ATM cell flow profiles are shown in Figures 8.14(b, c).

At the ATM side, the resulting traffic profile is determined by the policing scheme employed as outlined below:

- *VBR*.1: In a dual leaky bucket scheme, PCR flow is applied to CLP 0, 1 cells, and the SCR flow is applied to CLP 0, 1 cells. The conversion parameters in interworking will mark the Frame Relay DE-bit frames as equivalent ATM CLP = 1 cells. The ATM cells after T/2

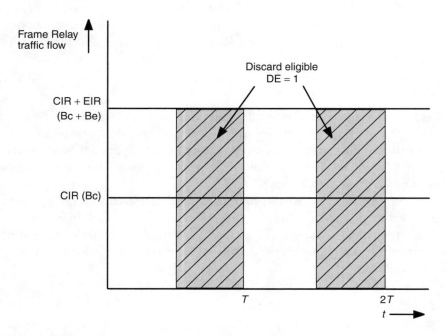

Figure 8.14(a) Frame Relay constant (Bc + Be) traffic rate.

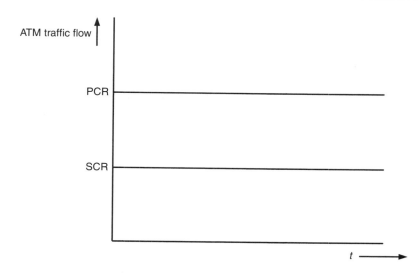

Figure 8.14(b) ATM PCR flow conformance.

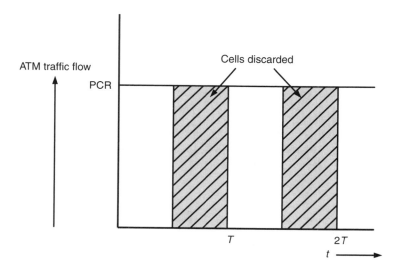

Figure 8.14(c) ATM SCR flow conformance for VBR.1.

will not conform to the SCR flow and will be dropped by the ATM policer, as shown in Figure 8.14(c).

- *VBR.2*: In this class of service, the SCR flow is applied to CLP = 0 cells only. Therefore, the CLP 1 cells will be ignored by the policer during the SCR flow check and, therefore, will generally pass through, as shown in Figure 8.14(d). However, in this case, the SCR flow will pass the GCRA (SCR, BT + CVDT) test.

- *VBR.3:* The main difference of this is that in VBR.3, nonconforming cells are tagged, whereas in VBR.2, nonconforming cells are dropped. Since, with the given SCR flow criterion the cells will be conforming, the resulting traffic profile will be identical to that for VBR.2.

8.4.4.3 Bursty Traffic

Assuming that the line rate is twice CIR + EIR, a worst-case scenario for Frame Relay traffic flow can be conceived as illustrated in Figure 8.15(a). In this case, the flow parameters have not changed from the previous case, but the line rate allows the incoming Frame Relay traffic to burst at full-line rate for $T/2$ period and idle for the following $T/2$ period. This pattern will satisfy the Frame Relay policer for the Bc criterion during the $T/4$ time period, and will conform to the Be test during the next $T/4$ period with DE bit set.

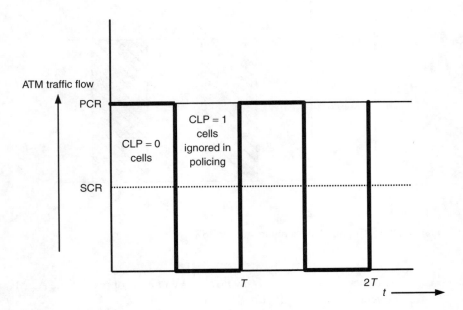

Figure 8.14(d) ATM SCR flow conformance for VBR.2 and VBR.3.

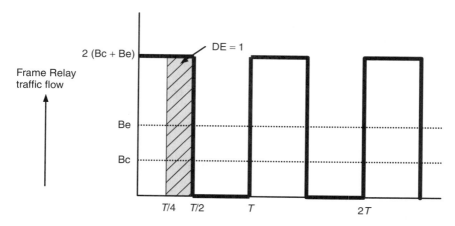

Figure 8.15(a) Bursty frame rate Frame Relay traffic profile.

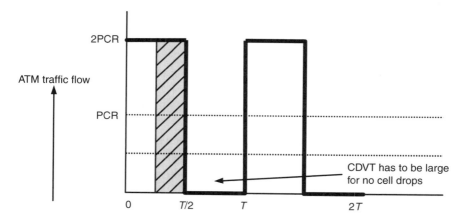

Figure 8.15(b) ATM PCR conformance.

The resulting ATM cell pattern is shown in Figure 8.15(b). The impact of this on the ATM side traffic policing is significant. The precise impact depends on class.

In VBR.1, conformance to ATM PCR flow is very likely to be violated unless the CVDT is set to a very large value determined by the following considerations. If we consider cell clumping, this scenario is similar to the derivation of the rules for determining SCR flow conformance, as illustrated in Figure 8.11.

Cells will be sent at twice the PCR rate for half the time period. There-fore, to be in conformant to the PCR flow, the CVDT must be set to a large value calculated from the following equation,

$$CDVT > (MBS–1)(1/PCR – 1/(2 \times PCR))$$

that is, $CDVT > (MBS–1) \times 1/(2 \times PCR)$

where MBS is the number of bits carried by the circuit during the whole of the time period (T).

However, since CLP0 + 1 is used in the SCR flow policing, at the mini-mum, half the total number of cells in the cell stream occupying the time period (0.25T to 0.5T) will fail the test, as shown in Figure 8.15(c). More will be lost, if the BT for the SCR flow is not defined to have a minimum value defined by

$$BT > (MBS–1)(1/SCR – 1/(4 \times SCR))$$

where MBS is the number of bits carried by the circuit during the first quar-ter of the time period.

In VBR.2 and VBR.3, conformance for PCR flow will be the same as that for VBR.1, with the requirement that CDVT has to be large to avoid cell drop. The SCR conformance requirement, however, is less stringent than that for VBR.1 because the already marked cells (CLP = 1) will not be policed in this group of services. Figure 8.15(d) illustrates SCR conformance for VBR.2 and VBR.3.

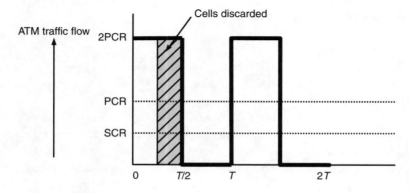

Figure 8.15(c)   ATM VBR.1 SCR flow conformance.

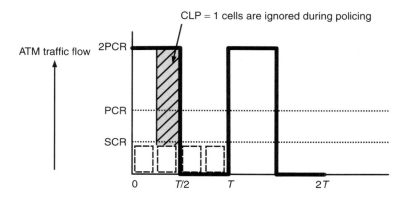

Figure 8.15(d) ATM VBR.2, VBR.3 SCR flow conformance.

8.5 QoS Enforcement in IP Networks

Key QoS-related concepts in IP networks are briefly reviewed here to facilitate proper understanding of QoS issues in ATM-IP interoperability scenarios.

IP networks have traditionally been implemented as best-effort networks, where all traffic streams are treated equally within available capacity of the network, resulting in delay and loss characteristics that vary over time, depending on the prevailing load and the state of the network. Network administrators depend on historical load patterns and the use of traffic measurements at critical points in network nodes to adequately overprovision link and node capacities to facilitate proper operation of their networks and avoid any debilitating congestion hot spots.

With the need to support multiservice networks, the IETF has been engaged in evolving strategies to provide end-to-end QoS in IP networks. This evolution can be best summarized by the following chronology of efforts in the IP-QoS area:

- *1995:* IntServ and RSVP mechanisms to implement microflow (per session) QoS handling. Scalability was found to be a major problem in faster links.

- *1998:* DiffServ efforts to define a limited number of QoS classes, traffic aggregation, and use of stateless core networks.

- *1999:* Attempts to use IntServ-RSVP scheme in enterprise and access networks and to use DiffServ in core networks.

Underlying QoS mechanisms for IP networks are similar to the concepts described for ATM networks and consist of the following:

- Traffic classification;
- Traffic conditioning, [e.g., policing, marking, shaping, or dropping];
- Queue management [e.g., random early discard (RED), weighted random early discard (WRED)];
- Queue scheduling [e.g., priority queuing (PQ), weighted round robin (WRR)].

A schematic diagram on the interaction of these QoS building blocks in a node is shown in Figure 8.16.

The application of the QoS components to obtain end-to-end QoS in IP networks depends on whether the router under consideration is part of the edge/access network or whether it is part of the core network. Routers have further been classified into provider edge (PE) and customer edge (CE) routers to identify the feature differences they are trying to accommodate. Further, the control plane implementation of end-to-end QoS (RSVP/IntServ or DiffServ) will impact the QoS components that are appropriate for different router types.

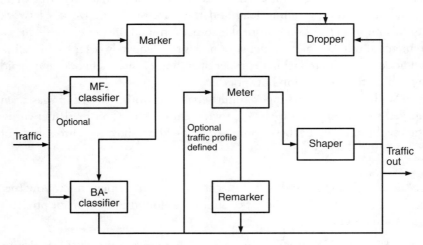

Key:
MF = multifield
BA = behavior aggregate

Figure 8.16 QoS building blocks.

Classification involves identification of packets for subsequent forwarding treatment. Packets are classified based on one or more fields in packet header (multifield classification), contents of packet payload, and input interface. Current implementations of classification algorithms are largely hardware-based, performed at wire speed, resulting in little performance impact.

IP QoS policing checks for conformance to a configured traffic profile. A policing algorithm, executed at the ingress of network, allows in-profile traffic into the network and marks, delays, or discards out-of-profile traffic. A token bucket algorithm that is modeled after the leaky bucket algorithm of ATM GCRA is often used for policing.

Shaping of traffic at the egress of networks forces the exiting traffic to conform to the traffic profile by removing the jitter introduced within the network. Shaping introduces latency to the traffic stream.

Queue scheduling and management is a key component that allows for differentiating IP traffic based on priority. Traditional queuing techniques, such as first in first out (FIFO), do not differentiate between services and can lead to network performance issues. Other methods have therefore been developed to differentiate between IP services.

8.6 Summary

Understanding the internal workings of policing algorithms that identify nonconformance cells is crucial for system designers building wireless networks. As seen, the algorithms operate in a simple stateless mode, continuously recomputing the criteria for accepting or rejecting the next cell based on its arrival time. Wireless networks inherently introduce jitter, clumping, and delay to ATM cells. The impact of such changes in the temporal characteristics of cell flow on traffic parameters needs to be quantified so that policers, if installed at the downstream end, can tolerate the changes in traffic parameters.

Further, since policers have to monitor typically hundreds to thousands of user connections individually in a node, the algorithms have been adopted for easy implementation in hardware. The ATM devices can be configured with policing parameters on a per-connection basis to implement nonintrusive policing at wire speed.

Selected Bibliography

ATM Forum Technical Committee, "Traffic Management Specification Version 4.1," af-tm-0121.000, March 1999.

Giroux, N., and S. Ganti, *QoS in ATM Networks*, Upper Saddle River, NJ: Prentice Hall, 1999.

Huang, A., C. Moreland, and I. Wright, "Advanced Traffic Management for Multiservice Networks," White Paper, Network Equipment Technologies, 1998.

ITU-Telecommunication Standardization Sector, "B-ISDN ATM Layer Cell Transfer Performance," Recommendation I.356, October 1996.

ITU-Telecommunication Standardization Sector, "Traffic Control and Congestion Control in B-ISDN," Recommendation I.371, May 1996.

McDysan, D., *QoS and Traffic Management in IP and ATM Networks*, New York: McGraw Hill, 2000.

9

Traffic-Management Issues in Wireless Internetworking Schemes

9.1 Introduction

Standardization bodies are intensively working on establishing end-to-end QoS in multiservice networks. In addition to the complex issues being addressed by the DiffServ and IntServ models being applied to IP networks, other common protocols such as Frame Relay and ATM interworking arrangements add further challenges to guarantee QoS in a heterogeneous network. Figure 9.1 illustrates the architecture of a network that will be used as an example for discussing QoS issues in a multiprotocol environment with LMDS wireless segments.

Defining service categories that are based on expected QoS is key to traffic management. Capacity calculation used in connection admission algorithms uses service categories to determine whether a new connection request is to be allowed or rejected. This chapter attempts to identify and illustrate how a wireless segment impacts the ability to ensure end-to-end QoS for a connection. As ATM has a well-defined QoS framework, the QoS issues are illustrated using a wireless link carrying ATM cells. Additional impact occurs when the source of the wireless link is derived from an interworking arrangement where a variable-length frame-oriented protocol such as Frame Relay is converted to fixed-length ATM cells.

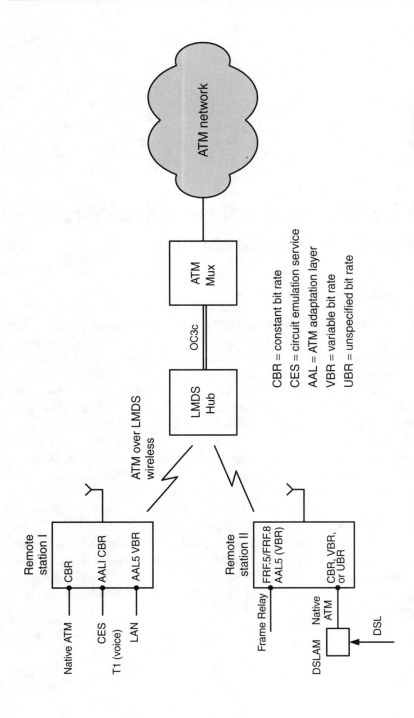

Figure 9.1 Typical LMDS configuration to illustrate QoS challenges.

9.2   Common Traffic-Management Issues

Internetworking arrangements and end-to-end connections in a network that includes wireless segments present several service provisioning issues related to QoS. Configuration of network elements and allocating network resources to provide guaranteed QoS service is fundamental to operators of commercial networks whose viability depends on extracting maximum capacity from the network. Factors that impact QoS in any heterogeneous network supporting multiple protocols are outlined in the following sections.

9.2.1   Variable-Length PDUs and Fixed-Length Cells

In Frame Relay to ATM interworking, or in IP to ATM interworking, variable-length frames or packets are converted into fixed-length ATM cells. Invariably, these interworking arrangements require the contents of only one frame to be carried in an ATM cell, thereby introducing inefficiency due to partially filled ATM cells. Further, additional headers, trailers, or both are required at the far end to reassemble the frames/packets. These overheads modify the traffic parameters across the interworking domains, no matter what physical links are involved.

9.2.2   Forced Traffic Shaping Due to Wireless Protocols

The traffic profile at the ingress end of wireless link will be influenced by the characteristics of the wireless protocol, resulting in a modified traffic profile at the egress end. The extent of the profile transformation will generally depend on the following factors:

- The wireless protocol is TDM based, and there will be assigned slots in the TDM frames that carry cells. In a lightly loaded system, not all the slots will be configured to carry traffic.

- The queuing strategy implemented at the ingress end determines if a cell from a specific session is transported when a TDM slot becomes available.

- If a frame-to-cell transformation takes place at the ingress end in an interworking situation like AAL5, cells generated from a specific frame may tend to bunch together in the queue, delaying queue entry to cells from other competing sessions.

- When a cell is available to be transported, a corresponding time slot may not be available, increasing delay (or jitter).

Later sections in this chapter will attempt to quantify some of the above impacts in a typical LMDS network.

9.3   Impact of TDM-Based Wireless Protocol

In the cases where there is a direct interworking of Frame Relay and ATM, and there are no wireless protocol segments, the cell jitter problems due to line rates higher than the Bc, Be rates in Frame Relay can be accommodated by incorporating appropriate smoothing functions at the output end of the Frame Relay in the Frame Relay to ATM direction. However, this is not sufficient when a wireless link based on TDM is involved.

9.3.1   A Typical TDM Scheme

An analysis of ATM cell clumping due to an intervening wireless segment can provide us with some insight into setting the policing parameters at the egress end of the wireless link. It has to be noted, however, that a systemwide look at the cell/frame flow is essential, as suitable smoothers can be implemented (if additional delays can be accommodated) to compensate for the jitter caused by clumping.

For a hypothetical case, let us assume a DS3 capacity link, with the highest capacity slot carrying a cell representing an effective data rate of 64 Kbps. Let us also assume that there are some overhead slots, as shown in Figure 9.2(a). Typical modulation techniques used by products like Hughes Network Systems' AirReach product are QPSK, 16 QAM, and 64 QAM. The relative bit density supported by these three modulation schemes are in the ratio 1:2:3 shown in Figure 9.2(b). Also, typically a quad-cell format is employed to optimize cells per burst and minimize burst overhead. This format includes a preamble (guard bits, ramp-up, and synchronization pattern)

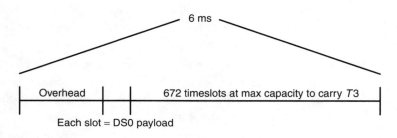

Figure 9.2(a) TDM configuration.

and trailer consisting of Reed-Solomon check bits and some spare bits. A sample configuration is illustrated in Figure 9.2(c).

9.3.2 ATM Cell Clumping and the Required CDVT

In the TDM scheme described, we will now illustrate a few possible ATM cell flow scenarios and how traffic policing will be affected. A specific constant rate cell flow is assumed at the input end of the wireless segment, and observations are made as to how these are impacted by different TDM slot assignments.

Typical cell capacity for different modulations

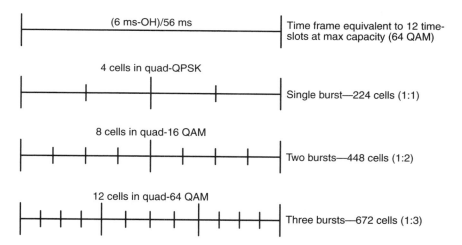

Figure 9.2(b) Modulation capacities of QPSK, 16 QAM, and 64 QAM in a TDM system.

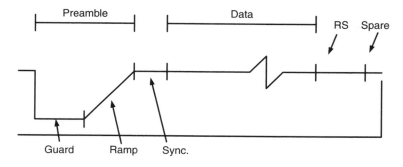

Figure 9.2(c) Burst configuration in a TDM system.

Case 1: A 64-Kbps constant Frame Relay payload with only one quad-slot assigned   Since a quad-cell scheme is used for the wireless protocol, a minimum four-cell slot assignment is made during the ATM connection establishment phase. If the input Frame Relay flow is at a CBR of 64 Kbps, the converted ATM cell rate will be 1 cell per TDM frame. However, the modulation/burst constraints impose a four contiguous cell assignment in the TDM frame, as shown in Figure 9.3 for the three different modulation cases. If we look at how cells are going to be transported in the TDM frame, the best possible scenario is that only one cell be waiting in the input queue when the four consecutive time slots arrive; therefore, only one cell per frame will be transported. However, in the worst case, at the ingress end of the wireless link, more than four cells have collected in the input queue and will be transmitted within the four slots available in the TDM frame.

When PCR (128 Kbps) is twice SCR (64 Kbps), we can derive the CDVT to be set in the ATM multiplexer at the egress end of the wireless link end so that the PCR flow is in conformance:

Quad-QPSK: $\text{CDVT} > (n-1)(6 \text{ ms} - 6/(56 \times 4))$

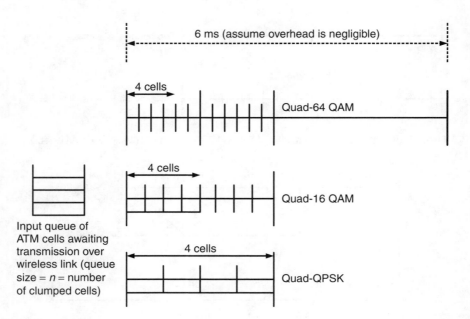

Figure 9.3 TDM slot assignments in three schemes for a 64-Kbps link.

where n is the maximum number of clumped cells. We find that

$$CDVT > (4-1) \times 6[1 - 1/224]$$
$$CDVT > 17.919 \text{ ms}$$

Quad-16 QAM: $\quad CDVT > (n-1)(6 \text{ ms} - 6/(56 \times 2 \times 4))$
$$CDVT > 17.959 \text{ ms}$$

Quad-64 QAM: $\quad CDVT > (n-1)(6 \text{ ms} - 6(56 \times 3 \times 4))$
$$CDVT > 17.973 \text{ ms}$$

These calculations illustrate that the CDVT is less affected by the modulation (cell density and, therefore, the peak possible cell rate) than by the availability of consecutive slots in the TDM frame.

Where cells from different ATM VP and VCs to the same hub are carried in a multiplexed form in the TDM time slots, cell behavior in one connection impacts the flow behavior in other connections.

Case 2: A 64-Kbps constant Frame Relay payload with two quad-slots assigned
When other sessions carried by the TDM link require a slot capacity of more than four slots (i.e., assume that there is another ATM VC that requires 256-Kbps bandwidth), then the slot assignment algorithm will likely assign the next quad-cell slots in the middle of the TDM frame, thereby minimizing the possible initial delay incurred for the end-to-end cell profiles. See Figure 9.4 for this illustration.

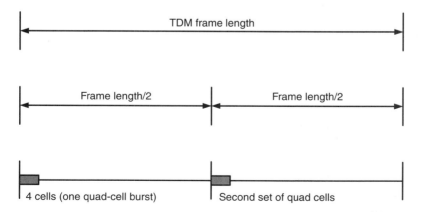

Figure 9.4 Two-slot assignments for two quad-cell bursts.

In this case, the 64-Kbps and the 256-Kbps sessions will be multiplexed on the available eight slots based on the cell availability from the incoming streams and any priority-based scheduling implemented in the input-queues servicing the wireless link. For CDVT calculations, the maximum clumping from the 64-Kbps stream will occur if more than four of the eight slots are occupied by the 64-Kbps stream cells.

It can be easily seen that the CDVT calculations shown for case 1 still will be applicable to case 2. The difference between the two cases is that since there are TDM slots available in the midstream of the TDM frame, the cells from the 64-Kbps stream, which missed the first quad-slots, instead of awaiting a full TDM frame time, will be able to find slots in the mid-TDM frame to carry them. This reduces the probability of cell clumping. It also has to be noted, however, that this scenario will be complicated by the 256-Kbps stream cells competing for time slots.

Case 3: A 64-Kbps constant Frame Relay payload with fully configured TDM frame   If all slots in the TDM frame are configured for ATM cell transmission, the worst possible case of cell clumping will arise under the following scenario. When the queues for all sessions except the 64-Kbps session are empty, cells from the 64-Kbps session can occupy as many of the adjacent slots as the number of available cells in the queue. Also note that these cells will not have accumulated if there were other available slots during previous time slots. Hence, this special case will only arise if cells from other multiplexed higher priority streams occupied the time slots while the 64-Kbps stream cells accumulated in its queue.

In a quad-QPSK modulation scheme, the equation for compliance of the CDVT will be

$$CDVT > (n-1)(6 - 6/(56 \times 4))$$
$$\text{that is, } CDVT > (n-1) \times 5.973.$$

Therefore, if there is a chance of n consecutive cells that can be clumped, the CDVT values to pass the PCR conformance test will be

4 cells   17.919 ms
5 cells   23.892 ms
6 cells   29.865 ms
7 cells ...

Similarly, CDVT values corresponding to quad-16 QAM and quad-64 QAM schemes can be derived using the following equations:

Quad-16 QAM: $CDVT > (n-1) \times 5.986$

Quad-64 QAM: $CDVT > (n-1) \times 5.991$

These values are only fractionally greater than those derived for the quad-QPSK scheme. The sensitiveness to CDVT with regards to ATM cell clumping is a common phenomenon in TDM-based wireless LMDS. Frame-based (AAL5) VBR connections are more susceptible to cell clumping than short payloads used in CBR connections such as CES. When adding new service connections, there will be CDVT violations due to cell clumping at the ingress of the ATM multiplexor (mux) if the configured CDVT is too small. The cell clumping will get worse with increased traffic loads. In order to avoid unnecessary cell drops, it is best to configure the CDVT value at the ATM mux to be much larger than typical ATM wire line configurations. As we can see from the previous calculations, the larger the CDVT, the more number of ATM cells will be allowed to arrive back-to-back at the ATM mux. Most hardware vendors allow CDVT values between 500 μs to a few seconds. A typical best-case value is 5,000 μs, and a worst-case value is 35 ms. If shaping is implemented at the hub station, then the best-case value would suffice as all traffic entering the ATM mux will be shaped.

9.4 Dynamic Bandwidth Allocation in LMDS

Dynamic bandwidth allocation (DBA) provides more efficient utilization of over-the-wireless link for bursty traffic flow between LMDS remote and hub stations. DBA allows for oversubscription of the network due to the bursty nature of LAN or other kinds of traffic. DBA allocates and deallocates band-width on demand for non-real-time applications such as Frame Relay, LAN, and native ATM services at the remote station. Without DBA, valuable bandwidth will be underutilized, as non-real-time traffic tends to be bursty in nature. It is impractical to allocate the entire wireless link for a 100 BaseT Ethernet LAN service.

At configuration time, air bursts are allocated for each service connection that requires guaranteed (committed) bandwidth. The bandwidth allocated depends on the ATM traffic parameters for the service connection. Thus, the total bandwidth configured does not change until a new service connection is added or an old service connection is deleted. The clear advantage of DBA is the increase in the rate of oversubscription in a given hub station sector. DBA allocates a minimum bandwidth based on SCR or MCR for non-real-time service connections.

In DBA, the remote LMDS station sends air burst allocation requests to the hub station when it requires additional bursts, as shown in Figure 9.5. The hub station responds with the allocation of the air burst that may be used in the uplink direction. In a PMP configuration, several remote stations can request bursts, and the hub station's air burst manager must resolve these requests by allocating bursts to each requester. Similarly, the remote station deallocates bursts by sending a deallocate request to the hub station. The deallocated air burst can then be used by another remote station if needed. Air bursts for the uplink belong to a common pool at the hub station for DBA purposes. This way, air bursts can be allocated/deallocated as needed by the hub station.

The criteria for the allocation and deallocation of air bursts in DBA is achieved by monitoring the frame queue depth of the AAL5 SAR at the remote station and setting threshold values for onset and abate conditions, as

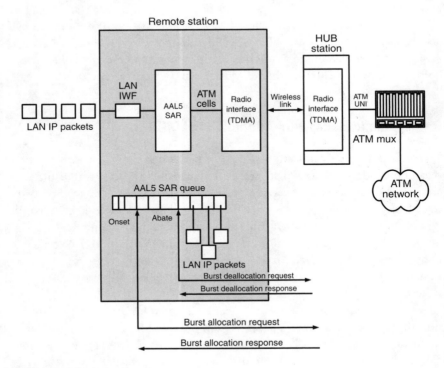

Figure 9.5 VBR service with DBA.

shown in Figure 9.5. When the queue depth exceeds the onset value, then additional air bursts are requested from the hub station. When the queue depth falls below the abate threshold, the remote station can de-allocate the air bursts that are not needed. There are always a minimum number of air bursts allocated (equivalent to SCR) when each service connection is added, thus guaranteeing the frame queue depth to be below the abate threshold when the traffic rate is equal to or below the SCR.

Clearly, we can see that in order to implement non-real-time services and make more efficient use of wireless bandwidth, there is a need to implement DBA between the remote station and hub station. This way, initially, the air bandwidth is allocated the equivalent of SCR, and air bursts are allocated and deallocated as needed when the traffic bursts above SCR up to PCR, as shown in Figure 9.6.

9.5 LMDS Application Examples

To conclude this discussion of traffic management in wireless interworking, we will now consider the functions of ATM traffic scheduling, shaping, policing, and statistical multiplexing as they apply to LMDS and discuss issues when implementing this technology in conjunction with a TDMA-based radio interface. The concepts are best illustrated with an example that shows how an end user will implement various types of services in a multimedia application.

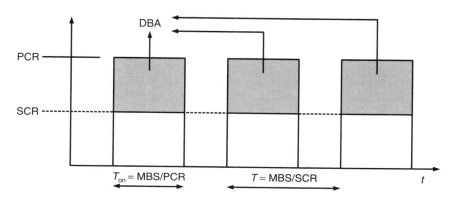

This periodic on-off source represents the nearly worst-case traffic pattern compliant to the traffic descriptor; the source emits cells at the peak rate (PCR) for a maximum duration of MBS/PCR

Figure 9.6 DBA for ATM rt-VBR and nrt-VBR services.

Figure 9.7 illustrates the role of ATM in LMDS where two remote stations and a hub station provide broadband access via an ATM backhaul network. Remote station A has three interfaces. The first connects the station to a video encoder using an OC3c physical interface. The second interface is a T1 connected to a PBX, which provides 64-Kbps PCM voice services. The third interface is an Ethernet LAN connected to an IP router. The operator at the network management center configures services for each connection as native ATM CBR for the video, CES (structured, no CAS) CBR for the PBX, and nrt-VBR for the LAN. The CAC at the Network Management Center verifies the traffic profile for each service connection before the configuration is allowed at the remote station A and hub station. The LMDS

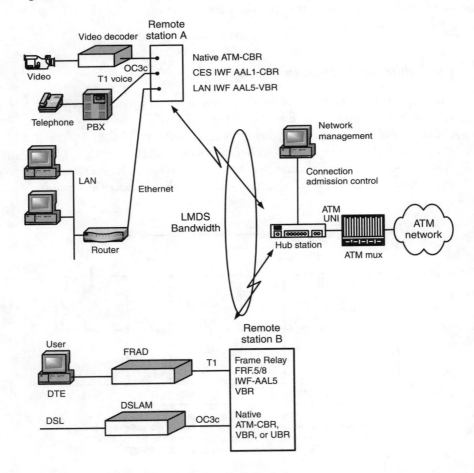

Figure 9.7 Example of an LMDS service configuration.

bandwidth for the three services at the remote station A and hub station is configured as follows:

- PCR of 28,302 cells/s (12 Mbps) for the video CBR connection. Note that the bandwidth of 12 Mbps is converted to ATM cells/s by dividing with (53 × 8), as 53 bytes makes an ATM cell that includes the cell header. LMDS bandwidth is allocated for the entire PCR.

- PCR of 4,107 cells/s for the 24xDS0 T1 (no partial fill, no CAS) CBR connection to the PBX. See Section 9.5.7 for the equations for the calculation of PCR. LMDS bandwidth is allocated for the entire PCR.

- SCR of 10 Mbps, PCR to the maximum air interface bandwidth of the LMDS (e.g., DS3 rate of 96,000 cells/s) system and MBS of 100 cells for the LAN nrt-VBR connection. This implies that the traffic is allowed to burst up to the maximum air bandwidth of the LMDS with maximum burst size of 100 cells for a duration of 1.04 ms (MBS/PCR), which is the burst tolerance (BT). Air bandwidth is allocated only for the equivalent of SCR. DBA allocates/deallocates air bandwidth as required when the traffic bursts above SCR for a duration given by BT.

In this example, the remote station B is connected to a T1 FRAD. It is also connected to a DSLAM via an OC3c physical interface. In this application, the AAL5 SAR may terminate a variety of data services (transported via Frame Relay over low-speed interfaces). This application is popular on ATM line interfaces because Frame Relay is a prevalent transport technology for enterprises. In fact, this application illustrates the most common use of SARs in WAN infrastructure equipment. The Frame Relay IWF may be either FRF.5 or FRF.8 at the remote station, as described in Chapter 6. The Frame Relay DLCI connection is mapped to an ATM connection using nrt-VBR. The DSL interface is provided by a DSLAM at the remote station B. The connection to the DSLAM can be configured as a native ATM CBR, rt-VBR, nrt-VBR, or UBR connection, depending on the DSL applications.

The LMDS bandwidth for the two services at remote station B is configured as follows:

- The Frame Relay DCLI connection is configured as Bc = 768 Kbps, Be = 768 Kbps, and CIR = 768 Kbps on the T1. The SCR, PCR, and MBS values are derived from the Frame Relay DLCI traffic parameters using equations (two GCRA characterization formula that provides EIR matching) given in Chapter 6. This works out as 5,052 cells/s for PCR, 2,526 per second for SCR, and 5,052 cells for MBS for a worst-case frame size of 38 bytes. The traffic for the Frame Relay connection is allowed to burst up to the PCR air bandwidth of 5,052 cells/s for a duration of 1 sec (BT = MBS/PCR). Air bandwidth is allocated only for the equivalent of 2,526 cells/s (SCR). DBA allocates/deallocates air bandwidth as required when the traffic bursts above SCR.

- The DSL connection is configured as an UBR-plus service in which the minimum desired cell rate (MDCR) and the PCR traffic parameters are relevant. The air bandwidth is allocated for MDCR. DBA allocates/deallocates air bandwidth as required when the traffic bursts above MDCR, up to PCR.

The hub station is connected to an ATM multiplexer, which is connected to the ATM backhaul network, from which subscribers can access a variety of networks such as PSTN, public and private Frame Relay or TDM networks, and the Internet. The ATM multiplexer and several hub stations are typically colocated at the hub site, as shown in this example.

Figure 9.8 illustrates the details of the ATM functions that are implemented at the remote station, hub station, and the ATM multiplexer for the various types of ATM service configurations included in this example. ATM traffic scheduling, shaping, and policing are done at different ingress and egress points in each of the segments, as indicated in Figure 9.8.

The hub station broadcasts in the downlink direction to all remote stations within the sector. The remote station will filter the traffic based on the ATM VP/VC connection or by the remote station MAC address. Finally, Figure 9.9 illustrates the LMDS bandwidth allocation at the time of configuration and DBA in a typical LMDS PMP configuration.

As illustrated in Figure 9.9, the following functions are implemented by the LMDS:

- Static bandwidth is allocated at configuration for the video, CES (PCM voice), and LAN services between remote station A and the hub station for the uplink.

- Depending on the traffic throughput, DBA allocates and deallocates air bursts for the LAN service at remote station A.

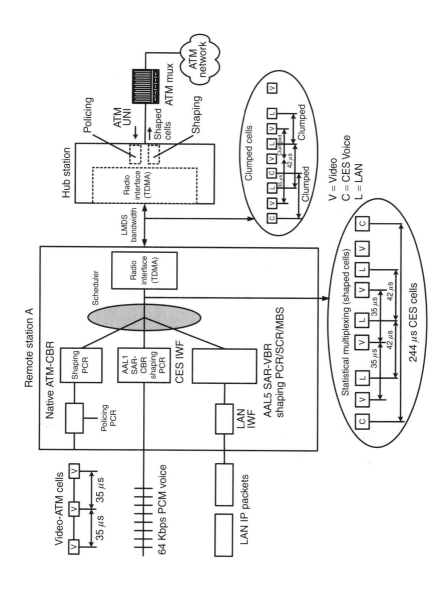

Figure 9.8 ATM statistical multiplexing and TDMA.

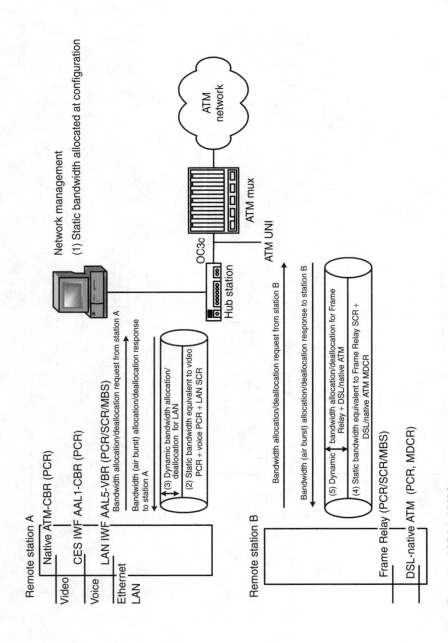

Figure 9.9 Example of DBA in LMDS.

- Static bandwidth is allocated at configuration for the Frame Relay and DSL-native ATM services between remote station B and the hub station for the uplink.

- Depending on the traffic throughput, DBA allocates and deallocates air bursts for the Frame Relay and DSL-native ATM services at remote station B.

9.5.1   ATM Traffic Scheduling and Impact on LMDS Example

Scheduling is a mechanism at the egress point of an ATM interface that controls the ATM QoS, manages proper bandwidth utilization, and provides fairness in allocating excess bandwidth. Traffic scheduling provides the necessary service differentiation to support a diverse range of QoS requirements for connections that belong to different traffic classes. Traffic is prioritized on the air interface by ATM service class, with highest priority given to CBR traffic, followed by rt-VBR, nrt-VBR, and UBR traffic. It is important to prioritize traffic in the uplink (remote station to hub station) direction for achieving efficient bandwidth utilization and fairness.

The QoS parameters are specified in terms of maximum end-to-end CTD, peak-to-peak CDV, CLR, or some combination of these depending on the ATM traffic class. While satisfying the QoS requirements, the scheduling mechanism must deliver the bandwidth committed as part of the traffic contract. The scheduling must ensure sufficient isolation between connections such that the traffic characteristics and QoS requirements of one connection do not adversely impact the bandwidth and QoS of another connection. It is also expected that the scheduler will provide fair share to connections of excess bandwidth whenever it is available.

In the LMDS example, ATM scheduling is performed at the following egress points, as indicated in Figure 9.8:

- At the remote station for outgoing traffic in the uplink direction;

- At the hub station for outgoing traffic toward the ATM multiplexer;

- At the hub station for traffic flow in the downlink direction (toward the remote station).

9.5.2   ATM Traffic Policing and Impact on LMDS Example

Traffic policing takes effect on the ingress of ATM connections. The policing function is used to monitor a connection to ensure that it conforms to the negotiated traffic contract, established at SVC call setup or PVC configuration. To perform policing, traffic parameters are monitored for each virtual connection using GCRA, as described in Chapter 8. The effect of policing function is one of the following:

- *Pass or tag*—Policing is not enabled for the connection, or the cell is accepted or tagged as conforming by the policing function.
- *Discard*—Policing is enabled for the connection, and the nonconforming cell is removed from the cell stream.

At the hub station in the downlink direction, traffic is queued according to ATM service class. At the remote station, traffic is shaped per connection. In the LMDS example, ATM policing is performed at the following ingress points, as indicated in Figure 9.8:

- At the remote station for native ATM traffic on the video OC3c interface;
- At the hub station for traffic coming from the ATM multiplexer.

9.5.3   ATM Traffic Shaping and Impact on LMDS Example

Traffic shaping at the remote station is required when cells have been clumped due to the TDMA air interface as the air bursts may not exactly correlate to the traffic bandwidth allocation by the ATM scheduler, as shown in Figure 9.8. For example, CBR traffic cells for each connection must be paced out at 1/PCR.

Shaping is performed at the remote station in accordance with the following traffic parameters for each service connection:

- Video connection is a CBR connection, and ATM cells are paced at 35 μs (1/PCR).
- CES connection (64-Kbps PCM voice) is a CBR connection, and ATM cells are paced at 244 μs (1/PCR).
- The LAN connection is a nrt-VBR connection, and ATM cells are paced at 42 μs (1/SCR). Any excess is paced at 1/PCR (10 μs) and MBS (100 cells).

In this example, SCR-shaped VBR enables the LAN traffic to burst at the air interface bit rate for a maximum burst size of up to 100 cells before shaping the uplink traffic to SCR. This enables traffic with a service-level agreement to meet the committed information rate.

In the LMDS example, ATM traffic shaping is performed at the following egress points, as indicated in Figure 9.8:

- At the remote station for outgoing traffic in the uplink direction;
- At the hub station for outgoing traffic toward the ATM multiplexer.

9.5.4   ATM Statistical Multiplexing and Impact on LMDS Example

ATM provides for statistical multiplexing, which attempts to exploit the on-off, bursty nature of many source types, as shown in Figure 9.8. The advantage of LMDS with ATM core architecture is that statistical multiplexing provides voice and data multiplexing over the same wireless interface. ATM cells of different connections are interleaved at the ATM interface on to the wireless TDMA interface. This means that ATM cells belonging to CBR, rt-VBR, and nrt-VBR connections are multiplexed over the same air bursts, as shown in Figure 9.8. In the example, ATM cells for the video CBR connection, CES 64-Kbps PCM voice connection, and the nrt-VBR LAN connection are interleaved at the remote station before transmission over the wireless TDMA interface.

9.5.5   Delay in TDMA-Based LMDS

LMDS based on TDMA architecture are frame (slot) based and differ from wire line media when it comes to transmission of broadband traffic using ATM over radio. The air bursts are allocated in time and are available to the transmitter as specific time slots. In a PMP configuration, air bursts are allocated for traffic in the uplink direction to a remote station when adding a service configuration. Similarly, air bursts are de-allocated when deleting a service connection and are reassigned to another remote station as needed. Air bursts are allocated and deallocated whenever services are added or deleted on that remote station.

The delays associated with transporting the data on the allocated bandwidth is dependent on the position of the allocated air bursts within the frame in a TDMA-based LMDS. Lower delay is achieved by allocating multiple bursts within the frame. For CBR, the delay should not be greater than 3 ms.

This slotted nature of TDMA creates problems for ATM. The ATM scheduler, in conjunction with the shaper at the remote station, transmits a cell in accordance with the traffic contract for the connection. If the air burst is not available in time, there could be a delay in sending the cell, in addition to the packetization delay for CBR connections. This delay reduces as more service connections are added, because additional air bursts are allocated for this remote station. For CBR connections, the CAC must allocate air bursts in such a way that the air burst spacing meets certain delay criteria for the connection. The goal must be to space the air bursts allocation to yield a worst-case delay of 3 ms. For nrt-VBR and UBR connections, which are bursty in nature, initially there can be buffer depletion due to air bursts not being available in time during high-bandwidth utilization periods. In this case, it is best to disable the ATM traffic shaping function at the remote station to avoid potential buffer depletion.

9.5.6   Effects on ATM Shaping in TDMA

The effects of shaping on nrt-VBR and UBR connections at the remote station vary, depending on the number of air bursts that get allocated for the remote station. In Section 9.5.5, we mentioned that, for nrt-VBR and UBR connections, it is best to disable the shaping function. When more connections are added, the possibility of air bursts being allocated back-to-back increases.

The SAR will output ATM cells back-to-back over the TDMA interface during heavy traffic loads if the shaping is disabled. This will violate the CDVT at the ingress side of the ATM multiplexer, as policing is normally done at this input for cells that are traveling in the uplink direction. This phenomenon is known as clumping; that is, the cells are arriving too fast at the ATM multiplexer. The shaping function at the hub station will ensure that cells will be paced at the correct cell rate toward the ATM multiplexer. The buffering at the hub stations must be adequate to handle clumped cells.

It is likely that cells arriving at the hub station from the remote station (uplink) over the TDMA radio interface will be clumped. The effect of clumping depends on the number of ATM cells transmitted back-to-back for each air burst (slots), which is related to the type of modulation (QPSK, 64 QAM) of the radio interface, as discussed previously in this chapter.

9.5.7   CES Service Interworking Issues with TDMA-Based Wireless

Consider the CES service that provides an end-to-end ATM connection (i.e., from the input ATM interface to the output ATM interface) shown in

Figure 9.10. Traffic shaping is performed on cells generated by the CES IWF and transported by the ATM network. The ATM CBR guarantees bandwidth and provides the best quality, but it does not free up bandwidth during periods of voice-band inactivity. CES support for voice is the most common method used today, since the ATM Forum's specification promotes vendor interoperability.

Each ATM VP/VC connection configured for a CES service uses a portion of the wireless bandwidth that is allocated between the remote station and the hub station. The PCR determines the amount of bandwidth within the TDM-based air interface that needs to be allocated to provide the service within the QoS requirements for the CES ATM connection. The traffic conversion takes into account the AAL1 overhead and CAS overhead (if used). The PCR calculation also depends on whether the ATM cell is partially filled. ATM Forum's "Circuit Emulation Service Interoperability Specification" details the equations for the equivalent PCR calculations that the CES IWF will allocate for the unstructured and structured CES ATM PVC connections. For example, in the DS1 unstructured service at a nominal bit rate of 1.544 Mbps in which the two IWFs involved are emulating a DS1 circuit supplied via a DSX-1 interface, the PCR required for AAL1 transport of 1,544-Kbps user data is 4,107 cells/s. The calculation of the PCR is based on the following equation:

$$4,107 \text{ cells/s} > (1.544 \times 10^6 \text{ bps} + 130 \text{ ppm})/$$
$$(47 \text{ AAL1 octets/cell} \times 8 \text{ bits/octet})$$

The CES unstructured service does not rely on any particular data format. Bits received from the service interface are packed into cells without regard to framing. Note that no particular alignment between any octets in the frames and octets in an ATM cell can be assumed with unstructured data transfer. However, correct bit ordering must be used. Considering the 376 contiguous bits that will be packed into the AAL1 SDU, the first bit received

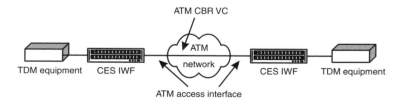

Figure 9.10 CES.

on the line is placed in the MSB of the first octet of the SDU. Placement then proceeds in order until the last bit is placed in the LSB of the 47th octet of the SDU.

The transmission delay over the radio, combined with packetization delay at the CES IWF, is a critical factor in TDM-based LMDS systems that transport voice using ATM. The packetization delay is the amount of time it takes to collect enough data to fill an ATM cell at the voice encoding rate. When using AAL1 at the CES IWF and the standard 64-Kbps rate for PCM, the resulting packetization delay is approximately 6 ms. The ATM network transports the cells containing the voice over AAL1. The CES IWF at the far end introduces a delay of 1 to 2 ms to account for variation in the arrival time of the ATM cells. If the packetization delay needs to be reduced, then it would be required to partially fill ATM cells using AAL1 rather than waiting for a full 46- or 47-byte payload before sending each cell. The disadvantage is reduced efficiency due to partial fill of the ATM cell payload. This reduces delay at the expense of higher cell rate. Partial fill is an optional feature in the CES IWF and, if available, the number of bytes to be sent in each cell can be set when the ATM VP/VC connection is established, whether through configuration of the PVC or by ATM UNI signaling for SVCs.

Most LMDS networks are built around ATM-based systems. The spacing of the allocation of air bursts due to the slotted nature of TDMA for the voice connection is critical for ATM CBR connections, as this factor alone can increase delay considerably. The air burst manager algorithm used for the allocation of air bursts at the LMDS Network Management Center must guarantee the spacing of the air bursts for each connection does not exceed 3 ms. ATM was not designed to propagate over radio, hence the wireless interface also needs to ensure that cells are delivered to and from the remote station and hub station without corruption. Any corruption to the 5-byte ATM cell header will result in the cell being dropped at the ingress point of the ATM interface. The FEC mechanism of the air link protocol will determine if the cell is corrupted and discard the cell before being forwarded to the ATM interface. The measurement of bit error rate (BER) between the remote station (downlink) and hub station (uplink) determines if the air link is good enough for the transmission of voice using CES. Generally, the BER should be better than 10^{-6} for the transmission of voice over LMDS. The collection of transmission link statistics on BER within the LMDS radio allows reporting of link degradation to the network management center.

CES's advantages are the simplicity of implementation. The ATM network is used to provide virtual replacements for physical links in an existing network. Nevertheless, CES has two limitations. It is unable to provide any statistical multiplexing. It does not differentiate between idle and active time slots, meaning that all idle traffic is carried. Therefore, CES voice transport consumes 10% more bandwidth than would be required to transfer the same voice traffic over leased circuits. CES is also often implemented as point-to-point service, providing the transport of the contents of one physical network interface to another physical network interface. This can prevent the implementation of some network topologies and can result in increased network cost. Multipoint voice services, such as voice conferencing or voice broadcasting, often require more sophisticated network solutions.

The limitations in CES resulted in the development of a 1997 standard referred to as dynamic bandwidth CES (DBCES), as shown in Figure 9.11. The objective of this standard is to detect active or inactive time slots of a legacy TDM trunk from a PBX or multiplexer, thereby dropping the inactive time slot from the next ATM structure. This enables this bandwidth to be reused for other services like CBR, VBR, UBR, and ABR applications. The DBCES standard specifies the ATM bandwidth can be dynamically adjusted to the number of active TDM time slots. The DBCES standard supports the following functions:

- TDM time slot activity detection;
- DBA according to the active time slots.

The TDM time slot detection method is beyond the scope of the DBCES specification. The method can be based on CAS or CCS. CAS, also known as robbed bit signaling (RBS), involves monitoring the ABCD bits of the 64-Kbps time slots to find out the status of the time slot (on-hook/off-hook conditions).

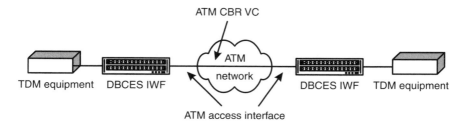

Figure 9.11 DBCES.

DBA involves dynamic sizing of the AAL1 structure that is used to carry the user data and signaling information. The AAL1 structure is sized to accommodate the maximum number of assigned time slots at configuration time. However, during the operation, the AAL1 structure may be dynamically adjusted up or down according to the number of active time slots that are detected in real time. The biggest difference between the original CES specification and DBCES is the payload and signaling structure. The AAL1 structures for DBCES are made up only of the active time slots. In the CES, the time slots may all be active, a mix of active and inactive, or all inactive. DBCES uses bandwidth more effectively than CES because it allocates bandwidth only to active time slots. The unused bandwidth can be temporarily assigned to another service type, such as UBR during lighter voice load. This temporary reassignment of bandwidth increases the effective bandwidth of the network.

Selected Bibliography

ATM Forum Technical Committee, "Circuit Emulation Service Interoperability Specification Version 2.0," af-vtoa-0078.000, January 1997.

ATM Forum Technical Committee, "Specification of (DBCES) Dynamic Bandwidth Utilization—in 64-Kbps Time Slot Trunking Over ATM—Using CES," af-vtoa-0085.000, July 1997.

ATM Forum Technical Committee, "Traffic Management Specification Version 4.1, af-tm-0121.000, March 1999

O'Leary, D., "Frame Relay/ATM PVC Network Interworking Implementation Agreement," FRF.5, December 20, 1994.

O'Leary, D., "Frame Relay/ATM PVC Service Interworking Implementation Agreement," FRF.8, April 14, 1995.

Technical Committee, "BISDN Intercarrier Interface (B-ICI) Specification Version 2.0 (Integrated)," af-bici-0013.003, December 1995.

10

Impact of Emerging Technologies on ATM Interworking

10.1 Current Networks and Technology

This book has covered the current state of heterogeneous networks, which include air-segments supported by technologies such as LMDS, in some depth. In addition to LMDS, MMDS, and other fixed wireless segments, other mobile wireless networks such as GPRS and the evolving 3G UMTS will be part of the heterogeneous networking landscape in the foreseeable future.

As we have shown, interworking functions between different access and transport protocols like Frame Relay, ATM, and IP present provisioning and management challenges. In addition to translating the different traffic-management parameters across different protocol domains, providing end-to-end QoS has been shown to present formidable issues for network service providers. Furthermore, if an air-segment such as an LMDS link is part of a user end-to-end session, then guaranteeing QoS across multiprotocol domains and air links has been shown to be an even more complex problem. Confronting this problem will involve intercarrier arrangements with partici-pating bandwidth brokering protocols, which are currently subjects of study in IETF.

Several new technologies are making rapid inroads in the networking area and are threatening to revolutionize data networking further. A list of evolving technologies that will soon impact bandwidth provisioning schemes, QoS negotiation, and delivery mechanisms are outlined in the sections below.

10.2   Evolving Network Technologies

Interworking and QoS issues will become more and more challenging in the future due to the advances in and adoption of new technologies. Inter-carrier arrangements related to QoS will become key to delivering service-level agreements that provide users with guaranteed QoS levels in heterogeneous networks that rely on multiprotocol and wireless segments. Sections 10.2.1 to 10.2.4 summarize selected key technologies that will increasingly be integrated in future multiservice networks.

10.2.1   DiffServ Model of QoS in IP Networks

Best-effort delivery in traditional IP networks is progressively being replaced by QoS enabled segments, modeled on the DiffServ standard.

As we have previously shown, service provider networks employ SONET/SDH leased lines, ATM, Frame Relay, and IP-based technologies.

Layer 2 technologies like ATM and Frame Relay include traffic control functionalities. However, Ethernet and packet over SONET (POS) do not provide layer 2 traffic control and, therefore, require layer 3 traffic control to achieve end-to-end. In general, for flexible management of different types of services, a combination of both layer 2 and layer 3 traffic-control functionalities are necessary.

In the DiffServ model, the ToS byte in the IP packet header, called a DSCP, is used for QoS marking. Traffic is classified at the edges of the network, and each packet is assigned to a per-hop behavior (PHB) group associated with a DSCP. Figure 10.1 shows the details of the DSCP format and the action of the routers at the edge and at the core networks. For backward compatibility with the IP precedence field (in the ToS byte), the RFC 2575 recommends the higher-order 3 bits to represent class selector code points and an all-0 code point to denote the default PHB value for the Internet.

PHBs defined by IETF determine forwarding treatment necessary to achieve the QoS for the component sessions of the integrated aggregate flows. Standard PHBs include best effort (BE), expedited forwarding (EF),

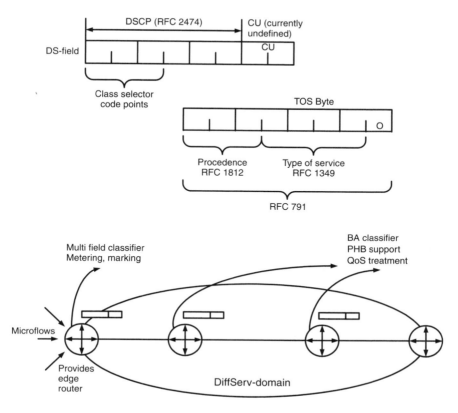

Figure 10.1 DSCP format and use of DSCP in edge/core routers.

and a set of assured forwarding (AF_{xy}) classes, where x defines AF class and y defines the drop preference (e.g., high/medium/low).

DiffServ DSCPs can be mapped to ATM service classes. An example mapping is shown in Table 10.1.

In a multicarrier network scenario, how the QoS for each segment of an end-to-end session is obtained from supporting networks is a topic of active study with different IETF work groups. Bandwidth brokering (BB), which provides traffic-control information to different networks, is a key aspect of these studies. The standards that emerge from these studies will impact the QoS issues in radio-frequency (RF) segments such as LMDS and MMDS. Policing and shaping of traffic generated from or destined for 2.5G and 3G wireless networks also will be impacted by the protocols adopted for bandwidth brokering in such multicarrier network environments.

Table 10.1
DiffServ PHB Mapping to ATM Service Classes

DiffServ PHB	ATM Service Class
EF	CBR
AF1	rt-VBR
AF2	nrt-VBR
AF3	GFR, ABR
AF4/BE	UBR

10.2.2 MPLS

MPLS has emerged as a versatile technology to address issues related to the speed of packet forwarding, scalability, QoS management, and traffic engineering in current IP-centric networks.

MPLS is essentially an advanced form of packet forwarding, where conventional longest address match forwarding is replaced with a more efficient label-swapping algorithm. Label is a short fixed-length value carried in a packet's header to identify an FEC, which is a set of packets that are forwarded over the same path [label-switched path (LSP)] through the network. Labels can be embedded in the link layer header (ATM VCI/VPI or Frame Relay DLCI) or can be inserted as a shim label between the layer 2 and layer 3 headers, as shown in Figure 10.2(a, b). Since MPLS can exist as an

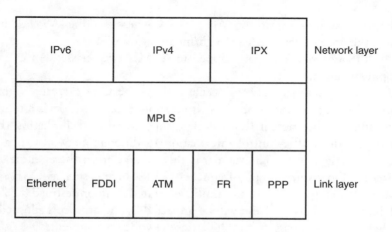

Figure 10.2(a) MPLS compatible protocols.

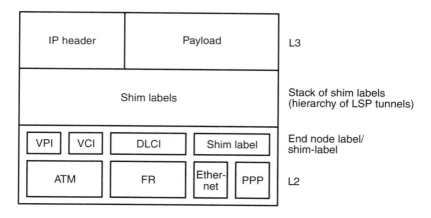

Figure 10.2(b) Shim labels for MPLS.

intermediate layer between IP (layer 3) and ATM/Frame Relay (layer 2), MPLS allows more efficient transfer of IP packets over Frame Relay and ATM networks.

The control component includes all conventional routing protocols (such as OSPF and BGP-4), and labels are distributed by piggybacking on a routing protocol or using a label distribution protocol (LDP). The label switching router (LSR) must create bindings between labels and FECs and distribute these bindings to other LSRs.

An LSP is functionally equivalent to a virtual circuit, as it establishes a fixed path between ingress LSR and egress-LSR for all packets assigned to an FEC. Service providers can establish customized LSPs to minimize the number of hops, force traffic across certain nodes, and for other reasons. Figure 10.2(c) illustrates a packet traversing an LSP.

MPLS supports IP-QoS by relying on platforms such as ATM-LSRs to provide guaranteed bandwidth LSPs. The MPLS label contains a 3-bit field labeled *EXP* (for experimental), which carries the marking for QoS classes. Since MPLS was defined before DSCP, the QoS bit field in MPLS took the form of the 3-bit IP precedence field to make the MPLS label width as small as possible. We see that when the number of QoS classes in the IP is less than 8, then the EXP field is sufficient to carry out the QoS marking in the MPLS network, and the LSPs are called E-LSP. When the IP network supports more than 8 PHBs, an alternative form of marking using the label itself, called L-LSP, is used. In addition, L-LSP field is used when the MPLS label is carried by an underlying ATM network (i.e., where there are no bits for marking QoS).

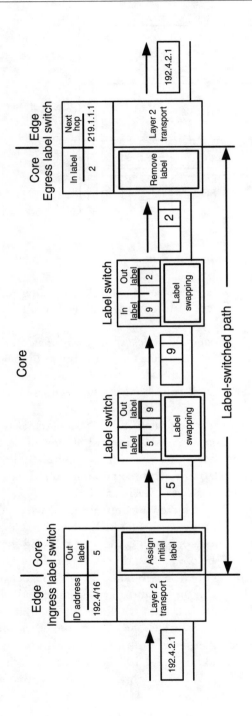

Figure 10.2(c) Label swapping on LSP.

It is seen that, as new technologies become widely adopted, challenging issues related to interworking and QoS negotiations are typically manifested. The requirements arising out of integrating fixed wireless (LMDS, MMDS) and mobile wireless (UMTS, GPRS) with wireline networks employing MPLS will pose formidable implementation challenges to network engineers.

10.2.3 Optical Switching

Fiber networks currently employ a SONET layer between the service layer and the optical layer, necessitating electrical-to-optical and optical-to-electrical conversion in nodes where add/drop multiplexing (ADM) and the cross-connect of SONET sessions occur. Migrating ADM, restoration, provisioning, and switching functions to the optical level is being planned by carriers using DWDM. This technology can support up to 240 wavelengths on a single fiber. Optical networking is likely to occur first at the core of the network, where optical pipes operate at highest capacity. Aggregation of traffic will continue to take place at the network edge. Optical switches employ nonblocking switch fabric that can scale to support increasing data rates.

Two different models are being currently studied for interoperability between the IP layer and the optical layer. The packet-routing view relegates end-to-end control intelligence to the IP layer (peer model), whereas the circuit-switched view makes the optical layer independently intelligent (signaled overlay model). Both these models are being studied by the IETF under GMPLS (generalized MPLS).

Enhancing MPLS to support a simplified IP-based control plane is key to the development of multicarrier interoperability models. Standardization has yet to address the requirements of IP transport over an optical layer. An MPLS-based control plane for optical networks containing multiple interconnected networks also needs standardization. Optical path provisioning is based on the well-established MPLS traffic-engineering framework and uses concepts from existing CR-LDP/RSVP (Constraint Routing-Label Distribution Protocol/Resource Reservation Protocol).

Wireless network designers must be aware of changing network technologies, especially with regards to traffic engineering that includes bandwidth allocation and QoS, so that wireless networks fit in seamlessly in evolving multicarrier, multiprotocol networks.

10.2.4 Mobile Networks

Third-generation (3G) mobile networks are conceived to combine high-speed mobile access with the already ubiquitous IP network services. 3G

UMTS, based on W-CDMA or CDMA-2000 technologies, will someday provide data rates up to 2 Mbps. 3G will be preceded by adoption of intermediate speed data services referred to as 2.5G technologies. 2.5G includes GPRS providing data rates from 56 Kbps to 114 Kbps. GPRS is based on the widely deployed GSM standard and, therefore, is essentially a software upgrade to existing BTS and BSC stations supporting today's GSM voice networks. Another 2.5G technology, enhanced data rates for global evolution (EDGE), provides speeds up to 384 Kbps by using all 8 slots in the GSM frame. EDGE allows the incumbent GSM operators (i.e., those without the UMTS spectrum) to closely approach speeds attained by future 3G UMTS networks.

The GPRS system architecture shown in Figure 10.3 illustrates the different disparate networks that can be associated in a user end-to-end session. The issues involved in guaranteeing QoS to an end user in this kind of network will challenge designers and integrators as this new generation of wireless systems becomes widely deployed and used.

To integrate GPRS into the traditional circuit-switched GSM architecture, two GPRS support nodes are added to the overall network architecture. The serving GPRS support node (SGSN) and gateway GPRS support node (GGSN), as shown in Figure 10.3, have been adopted in the general 3G architecture. SGSNs serve to connect the BSCs to the GPRS backbone. GGSNs serve as the gateway between the GPRS (radio) network and the external packet-data network.

It is seen that this architecture is inherently heterogeneous; therefore, interworking and tunneling (security) issues will manifest in the design of GGSNs. In addition, different operational domains and the multiple carriers involved in a typical end-to-end session will raise QoS issues that will engage system engineers for several more years.

QoS in 3G networks is defined in Third-Generation Partnership Project (3GPP) Forum documents. An end-to-end connection in Figure 10.3 can be laid out, as shown in Figure 10.4, to illustrate the QoS framework in the evolving 2.5G/3G wireless networks.

The mobile station negotiates the required QoS with SGSN, which relays this information to the GGSN when setting up the tunnel. The GGSN can accept, downgrade, or reject this request. On session setup, the GGSN maps the radio access network (RAN) side QoS to the IP-side QoS (e.g., by using equivalent DSCP values) to provide the session the required end-to-end QoS.

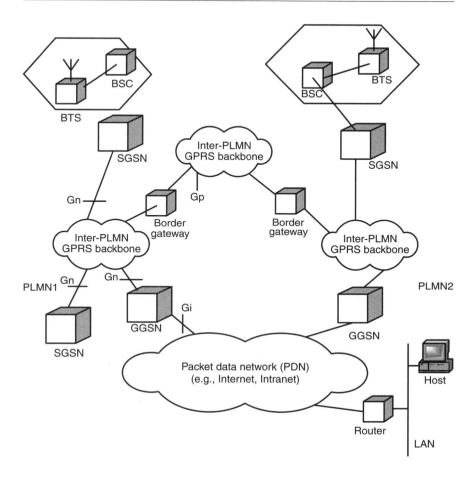

Figure 10.3 GPRS system architecture. (*From:* Bettstetter.)

10.3   Conclusion

Emerging technologies like DiffServ, MPLS, optical switching, and 2.5G/3G mobile networks will further increase the complexity of multiservices networks, creating new challenges for interworking and traffic management, including providing QoS guarantees for user sessions.

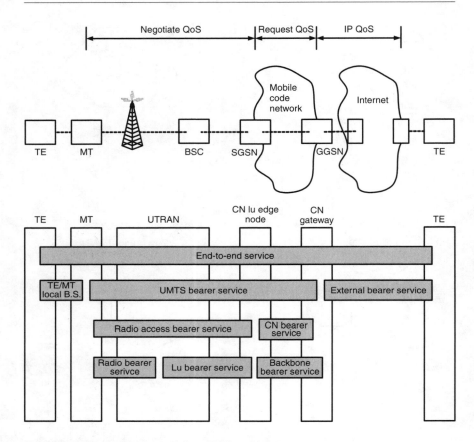

Figure 10.4 The QoS framework in a GPRS architecture.

Selected Bibliography

Blake, S., et al., "An Architecture for Differentiated Services," RFC 2475, December 1998.

Bala, K., "Internetworking Between IP and the Optical Layer," *Optical Networks Magazine*, May/June 2001, pp. 16–18.

Bettstetter, C., et al., "General Packet Radio Service GPRS: Architecture Protocols and Air Interface," *IEEE Communications Surveys* Vol. 2, No. 3 1999, pp. 2–14.

Davie, B., and Y. Rekhter, *MPLS—Technology and Applications* San Francisco, CA: Morgan Kaufmann, 2000.

Nichols, K., et al., "Definition of Differentiated Services Field in the IPv4 and IPv6 Headers," RFC 2474, December 1998.

Elloumi, O., "Providing QoS for Enterprise Networks Connected to Public QoS-Enabled IP Networks," White Paper, Alcatel, San Jose, CA, June 2000.

Elloumi, O., J. Perot, and M. Aissaoui, "Traffic Control for Optimal QoS in Next-Generation IP Networks," White Paper, Alcatel, San Jose, CA, July 2000.

Braden, K., et al., "Integrated Services in the Internet Architecture: An Overview," RFC 1633, June 1994.

"Interoperation of RSVP/IntServ and DiffServ Networks," Internet Draft.

Semeria, C., "Multiprotocol Label Switching," White Paper, Juniper Networks, 1999, at www.juniper.net/technet/techpapers/200001.html.

Acronyms

3G   third generation

3GPP   Third-Generation Partnership Project

AAL   ATM adaptation layer

ABR   available bit rate

ADM   add/drop multiplexing

ADSL   asymmetric digital subscriber line

AF   assured forwarding

AH   authentication header

AIS   alarm indication signal

AM   amplitude modulation

AMP   analog mobile phones

ANSI   American National Standards Interface

APPN   advanced peer-to-peer networking

AR   access line rate

ARP: Address Resolution Protocol

ARTT   Advanced Radio Telecommunications

AS   autonomous system

ATM   asynchronous transfer mode

ATMARP   ATM Address Resolution Protocol

B   bearer

BA   behavior aggregate

BB   bandwidth brokering

Bc   committed burst size

Be   excess burst size

BE   best effort

BECN   backward explicit congestion notification

BER   bit error rate

BGP   Border Gateway Protocol

B-ICI   B-ISDN intercarrier interface

B-ISUP   broadband-ISDN user part

BootP   Bootstrap Protocol

BPDU   bridge protocol data unit

BR   backward reporting

BRI   basic rate interface

BSC   base station controller

BT   burst tolerance

BTS   base terminal station

BUS   broadcast and unknown servers

CAC   connection admission control

CAS   channel associated signaling

CATV   cable television

CBF   CAS begin frame

CBR   constant bit rate

CCITT   Consulatif Internationale de Telegraphie et Telephonie

CCS   common channel signaling

CDMA   code division multiple access

CDPD   cellular digital packet data

CDV   cell delay variation

CDVT   cell delay variation tolerance

CE   customer edge

CER   cell error ratio

CES   circuit emulation service

C/I   carrier/interference ratio

CI   connection identifier

CIDR   classless interdomain routing

CIR   committed information rate

CLEC   competitive local exchange carrier

CLIP   classical IP

CLNP   Connectionless Network Protocol

CLP   cell loss priority

CLR   cell loss ratio

CMR   call misinsertion rate

CNR   carrier-to-noise

CO   central office

CPCS   common part convergence sublayer

CPE   customer premises equipment

C/R   command/response

CRC   cyclical redundancy check

CR-LDP/RSVP   Constraint Routing-Label Distribution Protocol/ Resource Reservation Protocol

CS   convergence sublayer

CSI   convergence sublayer indication

CTD   cell transfer delay

CU   currently undefined

D   data

DBA   dynamic bandwidth allocation

DBCES   dynamic bandwidth circuit emulation services

DCE   data communicating equipment

DCP   Data Compression Protocol

DE   discard eligibility

DEMS   digital electronic messaging service

DHCP   Dynamic Host Configuration Protocol

DiffServ   differentiated services

DLCI   data-link connection identifiers

DLSw   Data-Link Switching Protocol

DQDB   Distributed Queue Dual BUS standard

DSAP   destination service access point

DSCP   DiffServ code point

DSL   digital subscriber line

DSLAM   DSL access multiplexer or DSL access module

DS0   digital stream 0

DTE   data terminating equipment

DWDM   dense wavelength division multiplexing

EA   extended address

EDGE   enhanced data rates for global evolution

EF   expedited forwarding

EFCI   explicit forward congestion indication

EIGRP   Enhanced IGRP

EIR   excess information rate

ELAN   emulated local-area network

EOE   electro-opto-electronic

ES   end systems

ESF   extended superframe

ESIS   end system to intermediate system

ESP   encapsulating security payload

FCC   Federal Communications Commission

FCS   frame check sequence

FDD   frequency division duplex

FDDI   fiber distributed data interface

FDM   frequency division multiplexing

FDMA   frequency division multiple access

FEC   forward error correction

FECN   forward explicit congestion notification

FEP   front-end processor

FIFO   first in first out

FPM   forward performance monitoring

FR-CPE   Frame Relay customer premise equipment

FR-SSCS   Frame Relay service-specific convergence sublayer

FRAD Frame Relay access device

FRBS Frame Relay bearer service

FTP File Transfer Protocol

GCRA generic cell rate algorithms

GFC generic flow control

GFR guaranteed frame rate

GGSN gateway GPRS support node

GMPLS generalized multiprotocol label switching

GPRS General Packet Radio Services

GSM general special mobile

HDLC high-level data-link control

HDSL high-speed digital subscriber line

HEC header error check

HPR high-performance routing

HT hub terminal

HTTP Hypertext Transfer Protocol

IA implementation agreements

IAD integrated access device

IANA Internet Assigned Numbers Authority

ICMP Internet Control Message Protocol

ICW Internet call waiting

IDRP Interdomain Routing Protocol

IDSL ISDN digital subscriber line

IDU indoor unit

IEEE Institute of Electrical and Electronics Engineers

IETF Internet Engineering Task Force

IF intermediate frequency

IFL intermediate frequency link

IGMP Internet Group Multicast Protocol or Internet Gateway Message Protocol

IGP Interior Gateway Protocol

IGRP Interior Gateway Routing Protocol

IHL IP header length

ILEC incumbent local exchange carrier

ILMI integrated local management interface

I-Muxes inverse multiplexers

IN intelligent networks

InARP inverse ARP

InATMARP inverse ATM ARP

IntServ integrated services

IP Internet Protocol

IPX Internet Packet Exchange

ISA integrated services architecture

ISDN integrated services digital network

ISIS intermediate system to intermediate system

ISO International Standards Organization

ISP Internet service provider

ISUP ISDN user part

ITU International Telecommunications Union

IWF interworking function

IXC interexchange carriers

L2F Layer 2 Forwarding Protocol

L2TP   Layer 2 Tunneling Protocol

LAPB   Link Access Protocol-Balanced Mode

LAN   local-area network

LANE   LAN emulation

LCT   last conformance time

LDP   Label Distribution Protocol

LEC   LAN emulation clients

LECS   LE configuration servers

LE service   LAN emulation service

LES   LAN emulation servers

LIS   logical IP subnet

LIV   link integrity verification

LLC   logical link control

LLC/SNAP   logical control/subnetwork attachment point

LMCS   local multipoint communication service

LMDS   local multipoint distribution system

LMI   local management interface

LOS   loss of signal

LSB   least significant bit

LSP   label-switched path

LSR   label-switching router

MAC   Medium Access Control

MAN   metropolitan-area networks

MARS   multicast address resolution server

MBS   maximum burst size

MCR   minimum cell rate

MCS   multicast server

MDCR   minimum desired cell rate

MF   multifield

MFS   maximum frame size

MG   media gateway

MGC   media gateway control

MMDS   multichannel multipoint distribution system

MPC   MPOA clients

MPLmS   multiprotocol lamda switching

MPLS   multiprotocol label switching

MPOA   multiprotocol over ATM standards

MPS   MPOA servers

MSB   most significant bit

MSPS   megasymbols per second

MTU   multitenant units

MUX   multiplexor

NBMA   nonbroadcast multiaccess

NDIS   network driver interface specification

NetBIOS   network basic input/output system

NFS   network file server

NHC   next-hop resolution clients

NHRP   Next-Hop Resolution Protocol

NHS   NHRP server

NIC   network interface cards

NLPID   network level protocol ID

NNI   network-to-network interface

NSAP   network service access point

nrt-VBR   non-real-time variable bit rate

OAM   operations administration and maintenance or operation and management

OC   optical carrier

ODI   open data-link interface

ODSI   optical domain service interconnect

ODU   outdoor unit

OIF   optical internetworking forum

OSI   Open System Interconnection

OSPF   Open Shortest Path First

OUI   organizationally unique identifier

P2P-CDV   peak-to-peak cell delay variation

PBX   private branch exchange

PCM   pulse code modulated

PCR   peak cell rate

PDH   Plesiochronous Digital Hierarchy

PDU   protocol data unit

PE   provider edge

PHB   per-hop behavior

PID   protocol identifier

PINT   PSTN/Internet interfaces

PLCP   Physical Layer Convergence Protocol

PM   performance monitoring or physical medium

PMD   physical medium dependent

PMP   point-to-multipoint

POS packet over SONET

POTS plain old telephone service

PPP Point-to-Point Protocol

PPTP Point-to-Point Tunneling Protocol

PQ priority queuing

PRS primary reference source

PSK phase shift

PSTN public-switched telephone network

PTI payload type identifier

PVC permanent virtual circuit

QAM quadrature amplitude modulation

QPSK quadrature phase shift keying

QoS quality of service

RADSL rate adaptive digital subscriber line

RAN radio access network

RARP Reverse Address Resolution Protocol

RBOC regional Bell operating system

RBS robbed bit signaling

RDI remote defect indication

RED random early discard

RF radio frequency

RFC request for comments

RIP Routing Information Protocol

RM resource management

RPC Remote Procedure Call

RSVP Resource Reservation Protocol

RT remote terminal

RTS residual timestamp

rt-VBR real-time variable bit rate

SAP service access point

SAR segmentation and reassembly

SCES structured circuit emulation service

SCR sustained cell rate

SDH Synchronous Digital Hierarchy

SDLC synchronous data-link control

SDSL symmetric or single-line digital subscriber line

SDT structured data transfer

SDU service data unit

SEAL simple efficient adaptation layer

SECBR severely errored cell block ratio

SG signaling gateway

SGSN serving GPRS support node

SIP Session Integration Protocol

SMDS switched multimegabit data service

SN sequence number

SNA systems network architecture

SNAP subnetwork attachment point

SNMP Simple Network Management Protocol

SNP sequence number protection

SOHO small office/home office

SONET Synchronous Optical Network

SP signaling points

SPE synchronous payload envelope

SPIRITS PSTN/IN requesting Internet service

SRB Source-Route Bridging

SRT source root transparent

SRTS synchronous residual timestamp

SS7 Signaling System 7

SSAP source service access point

SSCS service-specific CS

SSP Switch-to-Switch Protocol

STM synchronous transport module

STP signaling transfer points or Spanning Tree Protocol

STS synchronous transport signal

SVC switched virtual circuit

TAT theoretical arrival time

TC transmission convergence

TCP Transmission Control Protocol

TDD time division duplex

TDM time division multiplexing

TDMA time division multiple access

TOS type of service

TTL time to live

UBR unspecified bit rate

UCES unstructured circuit emulation service

UDP User Datagram Protocol

UI user information

UMTS Universal Mobile Telecommunications Services

UNI   user network interface

UPC   usage parameter control

UU   user-to-user

VBR   variable bit rate

VC   virtual circuit or virtual channel

VCC   virtual channel connection

VCI   virtual channel identifier

VDSL   very high-bit-rate digital subscriber line

VOD   video on demand

VoFR   Voice over Frame Relay

VOIP   Voice over IP

VP   virtual path

VPC   virtual path connection

VPI   virtual path identifier

VT   virtual tributaries

WAP   Wireless Application Protocol

WAN   wide-area network

WDM   wave division multiplexing

WLL   wireless local loop

WRED   weighted random early discard

WRR   weighted round robin

About the Authors

Dr. M. Sreetharan was born in Jaffna (Thamil Eelam), Sri Lanka. He obtained his B.S. with highest honors from the University of Sri Lanka and taught electrical engineering for 2 years. He continued his postgraduate studies at the University of Manchester Institute of Science and Technology and held a research fellowship at Brunel University's (United Kingdom) Parallel Processing Group, obtaining his M.S. and Ph.D. in digital electronics in 1979 and 1982, respectively. He is the president of Performance Computing, Inc., which specializes in embedded systems development in mobile communication systems. Prior to his involvement with wireless networks, he worked on data communication systems and digital control systems. After working at Hughes Network Systems on the CDPD and point-to-multipoint product, he is currently a consultant at Megisto Systems, a Germantown, Maryland, startup, building high-speed mobile services platforms. He can be reached by e-mail at sree@webpci.com.

Mr. S. Subramaniam currently works at Hughes Network Systems, based in Germantown, Maryland. He obtained his B.S. in electronic engineering and M.S. in computer science from City University, London, England. He is a lead systems engineer for ATM, Frame Relay, and CES service-related development. He has worked in the Wireless Local Loop Group for several years at Hughes in the design and development of software applications. Prior to working at Hughes, he worked as a systems engineer at Aeronautical

Radio, Inc., on the design and development of packet switches for the air-lines- packet-data network and at Atlantic Research Corporation on implementing SS7, X.25, Frame Relay, and SNA protocols. At Standard Telephone and Cables in London, England, and Bell Telephones in Antwerp, Belgium, he worked on the design of the call processing and signaling software for digital class 5 switches. He can be reached by e-mail at ssubramaniam@hns.com.

Index

Digital Modulation Techniques, Fuqin Xiong

E-Commerce Systems Architecture and Applications,
Wasim E. Rajput

Engineering Internet QoS, Sanjay Jha and Mahbub Hassan

Error-Control Block Codes for Communications Engineers,
L. H. Charles Lee

FAX: Facsimile Technology and Systems, Third Edition,
Kenneth R. McConnell, Dennis Bodson, and Stephen Urban

Fundamentals of Network Security, John E. Canavan

Guide to ATM Systems and Technology, Mohammad A. Rahman

A Guide to the TCP/IP Protocol Suite, Floyd Wilder

*Information Superhighways Revisited: The Economics of
Multimedia,* Bruce Egan

*Integrated Broadband Networks: TCP/IP, ATM, SDH/SONET, and
WDM/Optics,* Byeong Gi Lee and Woojune Kim

Internet E-mail: Protocols, Standards, and Implementation,
Lawrence Hughes

Introduction to Telecommunications Network Engineering,
Tarmo Anttalainen

Introduction to Telephones and Telephone Systems, Third Edition,
A. Michael Noll

An Introduction to U.S. Telecommunications Law, Second Edition
Charles H. Kennedy

IP Convergence: The Next Revolution in Telecommunications,
Nathan J. Muller

The Law and Regulation of Telecommunications Carriers,
Henk Brands and Evan T. Leo

*Managing Internet-Driven Change in International
Telecommunications,* Rob Frieden

*Marketing Telecommunications Services: New Approaches for a
Changing Environment,* Karen G. Strouse

Telecommunications Deregulation and the Information Economy, Second Edition, James K. Shaw

Telemetry Systems Engineering, Frank Carden, Russell Jedlicka, and Robert Henry

Telephone Switching Systems, Richard A. Thompson

Understanding Modern Telecommunications and the Information Superhighway, John G. Nellist, and Elliott M. Gilbert

Understanding Networking Technology: Concepts, Terms, and Trends, Second Edition, Mark Norris

Videoconferencing and Videotelephony: Technology and Standards, Second Edition, Richard Schaphorst

Visual Telephony, Edward A. Daly and Kathleen J. Hansell

Wide-Area Data Network Performance Engineering, Robert G. Cole and Ravi Ramaswamy

Winning Telco Customers Using Marketing Databases, Rob Mattison

World-Class Telecommunications Service Development, Ellen P. Ward

For further information on these and other Artech House titles, including previously considered out-of-print books now available through our In-Print-Forever® (IPF®) program, contact:

Artech House
685 Canton Street
Norwood, MA 02062
Phone: 781-769-9750
Fax: 781-769-6334
e-mail: artech@artechhouse.com

Artech House
46 Gillingham Street
London SW1V 1AH UK
Phone: +44 (0)20 7596-8750
Fax: +44 (0)20 7630-0166
e-mail: artech-uk@artechhouse.com

Find us on the World Wide Web at:
www.artechhouse.com